REVELATIONS IN SCIENCE

"I want people to know things. I want them to understand. I want them to be able to run the world and enjoy the Universe."

—Isaac Asimov

Isaac Asimov has shared and imparted his wisdom and intelligence with readers for years. And in this twenty-sixth collection of essays from *The Magazine of Fantasy and Science Fiction* he demystifies and explains the universe and its secrets. America's most prolific author and a master explainer, Asimov's writings enlighten and educate causing *The New Scientist* to proclaim, "Asimov has probably awakened more people to the joy of science than any other author."

PINNACLE BOOKS HAS
SOMETHING FOR EVERYONE —

MAGICIANS, EXPLORERS, WITCHES AND CATS

THE HANDYMAN (377-3, $3.95/$4.95)
He is a magician who likes hands. He likes their comfortable shape and weight and size. He likes the portability of the hands once they are severed from the rest of the ponderous body. Detective Lanark must discover who The Handyman is before more handless bodies appear.

PASSAGE TO EDEN (538-5, $4.95/$5.95)
Set in a world of prehistoric beauty, here is the epic story of a courageous seafarer whose wanderings lead him to the ends of the old world — and to the discovery of a new world in the rugged, untamed wilderness of northwestern America.

BLACK BODY (505-9, $5.95/$6.95)
An extraordinary chronicle, this is the diary of a witch, a journal of the secrets of her race kept in return for not being burned for her "sin." It is the story of Alba, that rarest of creatures, a white witch: beautiful and able to walk in the human world undetected.

THE WHITE PUMA (532-6, $4.95/NCR)
The white puma has recognized the men who deprived him of his family. Now, like other predators before him, he has become a man-hater. This story is a fitting tribute to this magnificent animal that stands for all living creatures that have become, through man's carelessness, close to disappearing forever from the face of the earth.

ISAAC ASIMOV

THE SECRET
OF THE UNIVERSE

PINNACLE BOOKS
WINDSOR PUBLISHING CORP.

To Jill Roberts
and her fetching smile

PINNACLE BOOKS

are published by

Windsor Publishing Corp.
475 Park Avenue South
New York, NY 10016

First Pinnacle Books Printing: November 1992

Printed in the United States of America

Acknowledgments

The essays in this volume are reprinted from *The Magazine of Fantasy and Science Fiction*, having appeared in the indicated issues.

Contents

Introduction

Like all writers, I have troubles with reviews. I find myself unable to resist reading reviews of my books, and I find myself absolutely furious at anything less than an unalloyed rave.

I put the matter to my good friend, Lester del Rey, a man of infinite principle. He said:

"Isaac, never read reviews. If you find you must, then, at the first unpleasant adjective, tear it up and throw it away."

(Sigh.) How I wish I had his strength of mind.

A short time ago, I received a review of one of my latest collections of essays, of which this book is the most recent example. I would not ordinarily have seen it, for it appeared in a British newspaper. In Sri Lanka, however, my good friend, Arthur C. Clarke, was huddled over his fire, gumming his gruel, and *he* came across it. In a fever of agony over the possibility that I might not see it, he carefully clipped it out and mailed it to me.

Only the fact that I am a most forgiving man prevents me from hiring some ruffian to go to Sri Lanka and poison that gruel of his.

The review began: "This is a book that should never have been written."

Dear me! That activated del Rey's law and meant that the review should now be torn up and thrown away. I did that, but not before I scanned it quickly to see *why* it should never have been written. I gathered that what bothered the reviewer was that the book was a collection of miscellaneous essays on various subjects, hopping freely from one to another. Apparently, he thought that was unconscionable.

That's too bad. I presume he never heard of Charles Lamb, or of Michel de Montaigne, or even of Marcus Tullius Cicero.

Also, he was obviously ignorant of the fact that I have written nearly forty essay collections (mostly, but not quite entirely, on science) and that every one of them dealt with a miscellany of subjects among which I bounced about joyously.

Well, what does he expect? I have written, as part of this series and of others, a matter of six hundred essays or more, which have been collected in those forty books, and how can that be done unless I write on a wide variety of subjects. So the devil with the reviewer.

Here's another example. Recently, I published a large history of science entitled *Asimov's Chronology of Science and Discovery*. It was reviewed by an anonymous reviewer and the first phrase was: "The man who knows too much—" That activated del Rey's law and I tore up the review and threw it away.

But I kept thinking about the phrase. How can I "know too much"? What does the reviewer find offensive about my "knowing too much."

Actually, the remark is not true. I do happen to know a good deal about all the sciences and most of the humanities, but in the more prosaic fields of human endeavor I fall short.

For instance, I imagine that the reviewer can watch a football game or a basketball game on television and understand its every move. (It's all chaotic pandemonium to me.) I imagine he can chugalug a stein of beer like an expert. (I get faint at the odor.) I'll bet he can play poker like a dream and maybe can shoot a snappy game of craps. (I can't.)

It is precisely the abilities I lack and which the reviewer, perhaps, possesses in abundance that the American public finds admirable. To be familiar with the sciences and the humanities is (if I may trust some of the movies I have seen) the mark of a "nerd" and such a person is greeted with amusement and, it would almost seem, contempt. Most people, then, would feel I ought to be pitied for knowing *too little* about the things that really count.

And what are the things that *don't* count.

Well, a recent study has shown that American students have, in eighteen years, failed to advance their abilities to read and to write. A large fraction simply could not read or write (or think) back in 1972, and they cannot do any better in 1990.

The study blames this on (1) too much television, (2) not enough books, magazines, and newspapers in the home, and (3) not enough homework.

To cure this distressing situation, the study recommends that parents involve themselves more with what their children are doing.

Really? Surely, in a world that thinks I "know too much," parents are not going to suspect their children of knowing too little. As a matter of fact, I don't think most parents can read, write, or think any better than their kids can—or that they believe it is in any way important that they do so.

Incidentally, one part of the study seems to have been

that of asking children to answer questions that would test their values and their way of thinking. One child, in answer to the question "What would you consider a good job?" answered as follows:

"A good job is one in which I don't have to work, and get paid a lot of money."

When I heard that, I cheered and yelled and felt that he should be given an A+, for he had perfectly articulated the American dream of those who despise knowledge. What a politician that kid would have made.

So here I am—stuck. I want people to know things. I want them to understand. I want them to be able to run the world and enjoy the Universe. I've spent my whole life trying to explain the sciences and the humanities to them.

I reach several hundred thousand people, I know. It isn't much. The vast majority remain untouched, but every soul saved from the burning is a soul won for the light. Last week, believe it or not, I got a letter from Iran telling me that I was Iran's favorite science writer. If I can reach even Iran, how wonderful!

I don't really know too much, despite what the silly reviewer said. Actually, I don't know enough even about those subjects I am acquainted with. But what I do know I want others to know as well.

Part I
The Solar System

1
Smashing the Sky

When I was a very young lad—I couldn't have been more than seven at the time—I found a map of Greater New York. I had never seen a map before and I hadn't the faintest notion what it was.

There were curious shapes, and there were lines crossing here and there, and small print everywhere. However, as I studied it with puzzlement and curiosity, I came across some fairly large print which said BROOKLYN.

That excited me. After all, I knew that we lived in Brooklyn. The common conversation I heard from grown-ups around me made it clear that Brooklyn was home. I therefore looked eagerly for other words that made sense, and eventually, I discovered the names of streets that were familiar to me.

I remember the feeling of awe and gladness that I experienced. After all, my horizon was very close to my center, and I hadn't the faintest idea what lay beyond it. I therefore strongly suspected that the map I had which listed the streets I knew and many, many that I didn't know in dim faraway places like Queens and Manhattan must be a guide to the whole world—even the whole Universe.

I soon learned better, for in school, I eventually received geography books with maps covering more ground and I found that Brooklyn and even all the five boroughs of Greater New York are but an insignificant patch on an inconceivably larger world.

I felt the loss. For a brief period of time I'd thought I had a clear representation of everything there is, and that my knowledge was (at least potentially) total. To be introduced to misty distances plunged me into a rather fearful unknown.

But what I went through in my first decade, all of humanity went through over an enormous stretch of time. The night sky seemed, to humanity, a representation of everything there is, except for the solid Earth itself, and it was accepted as "all-there-is" with the same confidence that I accepted the map of New York as representing "all-there-is."

There were, of course, clouds, weather manifestations, shooting stars, and comets to be seen in the sky, but they were taken to be merely atmospheric manifestations. There were also the Sun, the Moon, Mercury, Venus, Mars, Jupiter, and Saturn, the seven "planets," which steadily shifted position relative to each other and to all the other stars. These were considered to exist between the Earth and the sky itself.

Aside from atmospheric manifestations and the seven planets, everything else was "sky." The sky in early times was viewed as a smoothly curved object, made of some thin solid material, and in the Bible it is called the "firmament," with that first syllable an indication that it was viewed as composed of something "firm" and solid. The word is Latin and it is a translation from the Greek *stereoma*, meaning a "solid dome," and that in turn is a translation from the Hebrew *rakia*, meaning a "thin, metallic sheet."

To the unsophisticated Biblical writers, the sky seemed a small and rather intimate semispherical solid dome that covered the flat Earth, coming down to meet it all around the horizon. Thus, when the Book of Revelation (written about A.D. 90) speaks of the destruction of Earth and sky, it says, "And the stars of heaven fell unto the earth . . . And the heaven departed as a scroll when it is rolled together" (Revelation 6:13–14). In other words, the thin metal out of which the sky was formed rolled up (spro-o-o-ing) and all the spangly little lights that had covered it fell off.

I would not be surprised if many millions of people on Earth had this same view of the sky and the stars at this very moment.

To the Biblical writers, the stars seemed innumerable, too, so when God wanted to tell Abram (later, Abraham) how many descendants he would have, "he brought him forth abroad, and said, Look now toward heaven, and tell [count] the stars, if thou be able to number them . . . So shall thy seed be." (Genesis 15:5).

Actually, however, the total number of stars visible to the normal eye is no more than 6,000. Only half of them are above the horizon at any one time, and the dimmer ones near the horizon are washed out by atmospheric effects even when the sky is completely clear. Therefore, the number of stars Abram could have seen was no more than about 2,500.

The Greek philosophers, as early as 500 B.C., were beginning to gather that Earth was a sphere. They viewed the sky as a large and perfect sphere at the center of which the spherical Earth was suspended. But even to them, the sky was a *solid* sphere, black by night and blue by day, and the stars, visible by night only, were tiny luminous dots attached to the solid sky.

The sphere, in their view, turned steadily about the Earth, completing one turn in twenty-four hours. The motion of the sky itself could not be seen, but the stars moved relative to the horizon and did so "all in a piece" as one would expect if they were all affixed to a solid dome. The stars were therefore called the "fixed stars," as opposed to the various "planets," which is from a Greek phrase meaning "wandering stars."

This was a grandly simple view, and prior to modern times, there were only a few people who tried to add anything significant to it. The Greek philosopher Democritus of Abdera (470–380 B.C.) maintained that the Milky Way consisted of numerous stars, too small to be seen individually, so that they melted into a kind of luminous fog.

The Milky Way was the only visible object affixed to the sky that did not resemble the tiny stars. Democritus's attempt to make it fit the other objects in the sky, and thus simplify the picture of the Universe, lacked evidence and was therefore not "compelling." It didn't force belief. It was much easier to believe that the Milky Way was, in actual fact, exactly what it appeared to be— a luminous fog. Besides, the Milky Way might be either a multitudinous crowd of stars or a luminous fog, but in either case, it did not disturb the picture of the sky as a solid sphere.

A much more revolutionary view was presented by a German scholar, Nicholas of Cusa (1401–1464). A book he published in 1440 claimed the following: Space is infinite and contains an infinite number of objects like our own Sun, each one shining over inhabited worlds, as our Sun does. The Suns are so far away that we see them only as tiny objects—stars—and we see only a few of them, since most are too far away to be seen at all.

It was a brilliant speculation that, for the first time, suggested that the sky is not a solid object, but that in its place there is merely infinite space. Again, there was no evidence and the view was not compelling in itself. People preferred to believe that the stars are exactly what they seem to be—little flecks of light affixed to a solid sky.

To be sure, a century and a half later, an Italian scholar, Giordano Bruno (1548–1600), picked up Nicholas's ideas and trumpeted them forth. He was a lot less tactful than Nicholas and times had grown a lot less tolerant. Therefore, whereas Nicholas died in his bed in the odor of sanctity as a cardinal of the Church, poor Bruno was burned at the stake.

In another direction altogether, a Greek astronomer, Aristarchus of Samos (310–230 B.C.), had maintained, about 260 B.C., that it made more sense to suppose that the planets, including Earth, revolve about the Sun (a "heliocentric" view), rather than to suppose that they, including the Sun, revolve about the Earth (a "geocentric" view). This was still a third noncompelling speculation that was not accepted because it went against appearances.

In 1543, a Polish astronomer, Nicholas Copernicus (1473–1543), picked up the Aristarchean speculation, and pointed out that while there was no compelling evidence for it, the heliocentric view made it easier to calculate the past and future motion of planets. That made it sound like a mathematical device and was not a truly compelling argument, so that for fifty years, the Copernican view failed to receive general acceptance.

Please notice, though, that substituting a heliocentric view for a geocentric one only affected Earth's immediate neighborhood. Even if the heliocentric view was adopted and the planets made up a "Solar system" as they circled "Sol" (the Latin word for the Sun), that did not necessarily affect the sphere of the sky. The sky might still be viewed as the same solid sphere it had always been thought to be, but one that had the Sun at its center rather than the Earth.

* * *

19

The heliocentric view became compelling, at last, when the Italian scientist Galileo Galilei (1564–1642) constructed a simple telescope and, in 1609, turned it on the sky.

To begin with, he found stars that were too dim to be seen by the unaided eye in every direction he looked, and that the Milky Way, as Democritus had maintained, was made up of innumerable very faint stars.

Then Galileo discovered the satellites of Jupiter and the phases of Venus, and that made it possible to adduce compelling arguments in favor of the heliocentric view. Galileo was himself forced to retract his arguments, under the threat of torture by the Inquisition, but astronomers, generally, were converted to heliocentrism.

This meant that the sky didn't really turn from east to west. It was the Earth that turned from west to east. Yet that didn't affect the nature of the sky either. It was still a solid sphere, black by night and blue by day, motionless rather than turning, and enclosing the Sun at its center rather than the Earth. It was still that old, familiar solid star-spangled structure.

Also, in 1609, the German astronomer Johann Kepler (1571–1630) maintained that planetary orbits are not circles, or combinations of circles, as everyone from the Greeks to Galileo had thought, but are ellipses instead. Kepler worked out the actual shape of the Solar system and was able to show the relative distance of the planets from Earth and from each other. But that didn't necessarily affect the nature of the sky either.

In 1672, the Italian-French astronomer Giovanni Domenico Cassini (1625–1673), working in Paris, was able, with the help of a French astronomer, Jean Richer (1630–1696), in French Guiana, to get the first reliable measurement indicating the scale of the Solar system. It turned out to be far, far larger than anyone had imagined. Cassini determined the orbit of Saturn, then the farthest known planet, to be something like 3 billion kilometers in diameter.

Yet that didn't necessarily change the nature of the sky either. It was a far huger sphere than had been thought, with a diameter not measured in mere miles as the Biblical writers had thought, or in thousands of miles as the early Greek philosophers had thought, or in millions of miles as Galileo and Kepler might have thought, but in *billions* of miles. Yet even though it had grown so monstrous, it might still be the same solid, star-spangled sphere.

In 1718, however, the English astronomer Edmund Halley (1656–1742) was checking the position of the various stars in order to prepare a new star map and he found that three bright stars—Sirius, Procyon, and Arcturus—had changed their positions (relative to the general run of dimmer stars) significantly since the time of the ancient Greeks.

From this he concluded that the stars are not "fixed" after all, but move relative to each other. It is just that they move so slowly that the changes in position become unmistakably only after they have been moving for hundreds, or even for thousands, of years.

But why should the stars move so slowly, when they do move, and why do most of the stars appear not to move at all? Halley surmised that the stars must be so enormously far away that their motions *seem* slow. In fact, only the very nearest of the stars would have visible motions even after many centuries. Stars that are still farther distant would show no visible movement even in the course of all the time that human beings have been observing the sky.

But if one assumes the stars are so enormously far away, surely one must turn to the speculations of Nicholas of Cusa nearly three centuries earlier. The stars must be Suns. Suppose the star Sirius, for instance, is as luminous as our Sun is. How far off would it have to be for its enormous luminosity to seem brighter than its apparent brightness as a star? Halley calculated what that distance would be and concluded (using our modern

21

system of measurement) that Sirius must be about 2 light-years away from us.

Halley was the first person to speak of the distance of the stars in terms of light-years and thus to suppose that if there were indeed such a thing as a solid sky sphere, it would have to be *trillions* of kilometers in diameter.

Yet Halley's arguments weren't compelling either. It might be that the stars are as dim as they are because they are comparatively close and *really* dim, and not because they are far-away Suns. It might be that a few of them drift very slowly across the solid surface of the sky because they are not completely fixed, but can shift a bit, very slowly. In that case, the sky would be much closer to Earth than Halley's speculations would lead one to believe.

Some direct and undisputable measurement of stellar distances was required, and astronomers knew how to do that, in theory. As the Earth swings from one side of its orbit to the other, it changes its position with respect to the surrounding stars by some 300 million kilometers. As it moves rightward in its orbit, the nearer stars seem to move leftward relative to the farther stars. This is "parallactic displacement."

The farther the stars, the tinier the parallactic displacement. If that displacement is large enough to be measured, however, then, knowing the diameter of Earth's orbit, you can calculate the distance of the star with a considerable degree of confidence.

When Copernicus first presented the heliocentric view, astronomers realized that the nearer stars ought to show parallactic displacement relative to the farther stars. From the fact that they don't, it was argued that Earth does not move. Copernicus, however, maintained that the stars are simply so distant that the parallactic displacement is immeasurably small—and Copernicus was right.

(Of course, whatever the distance of the stars, there would be no parallactic displacement if all were at the same distance, as they would have to be if they were affixed to the solid sphere of the sky.)

Eventually, telescopes were invented, and were steadily improved to the point where parallactic displacements that were far too small to have been seen earlier could be measured (if such displacements existed).

As it happened, they did, and the first to measure a parallactic displacement was the German astronomer Friedrich Wilhelm Bessel (1784–1846). In 1838, he announced the parallactic displacement of the star 61 Cygni, which turned out to be about 6 light-years away from us. In the next couple of years the distances of Alpha Centauri (4.3 light-years) and Vega (11 light-years) were determined by other astronomers.

It was clear that the stars, even the nearest, are farther away than even Halley had thought, and that they are not all at the same distance but are distributed through spaces so vast that one was drawn to Nicholas of Cusa's speculation of an infinite Universe.

So at last, the sky was smashed forever. There was no firmament, no solid sphere. Earth was surrounded only by space.

But now a new question arose. Are the stars really distributed though infinite space, equally in all directions, or is there some limit to their existence and do they form a finite grouping of some definite shape or other?

At first glance, the answer seems to be the one Nicholas of Cusa gave. If we look at the sky with a telescope, there are stars in every direction. The better the telescope, the more and dimmer (and therefore, presumably the more distant) the stars we see. It would seem that the Universe of stars forms a large spherical structure of enormous, and perhaps infinite, size.

There is only one catch to this, and that is that the distribution of the stars in the sky is not quite symmetrical. There is the Milky Way. If you look at any portion of the Milky Way, then you see an enormous number of very dim stars, a far greater number than you see in any portion of similar size anywhere else in the sky.

The first person to try to account for this asymmetry in the sky was a British astronomer, Thomas Wright (1711–1786). In 1750, after Halley's suggestion that the stars must be distributed through indefinite space, but before Bessel's (and others') confirmation of that, Wright worked up a mystical structure of the starry Universe as existing between two concentric spheres, with God and Heaven existing inside the smaller sphere. If you looked directly toward the inner or outer boundary of the starry realm, you would see relatively few stars. If you looked at right angles to that, through a long stretch of the star-filled in-between layer, you saw innumerable dim stars— the Milky Way.

The German scholar Immanuel Kant (1724–1804) picked up Wright's notion, in 1755, and either misunderstood it, or modified it. He got rid of the mystical elements and felt that the star system is lens-shaped. The Sun is at the center, and depending on whether one looked through the thin dimension or the thick one, one saw a scattering of the stars, or the Milky Way.

Both Wright and Kant were speculating, however, and did not make the kind of measurements that would make those speculations compelling. What was needed was supplied by the German-British astronomer William Herschel (1738–1822).

Beginning in 1785, Herschel actually counted the stars in the sky, not only those visible to the unaided eye, but those he could see in his telescope. It was impractical to try to count them all, so he marked out 683 small squares in the sky, distributing them as randomly as he could over the entire stretch of the sky visible to him. He then counted all the stars in each small square. (In this way, Herschel took a poll of the heavens and founded the science of "statistical astronomy.")

Herschel found that the number of stars in the various small squares increases steadily as one approaches the Milky Way, and that the number is smallest in directions at right angles to the plane of the Milky Way.

This made it clear, by actual measurements, that Kant was right and that the star system is indeed lens-shaped. What's more, by making some reasonable assumptions, Herschel came to some conclusions as to the size of the star system. He estimated that its long diameter extends for 8,000 light-years and its short diameter for 1,500 light-years, and that the whole contains 300 million stars.

This was a fearful underestimate but it was the first time that the Universe was spoken of in terms of tens of quadrillions of kilometers. Since the Milky Way is the result of looking through the star system the long way, that star system came to be called the "Galaxy," from the Greek phrase for the Milky Way.

Herschel may therefore be said to have discovered the Galaxy.

Since the Milky Way seemed to be more or less equally bright everywhere, and since the circle of the Milky Way seemed to divide the sky into two equal halves, it seemed reasonable to suppose that the Sun is at, or near, the center of the Galaxy.

Yet it is not fixed immovably at the center, and Herschel himself demonstrated that fact. Beginning in 1783, Herschel did his best to determine just how the stars are slowly changing position in the sky. He could do this much better than Halley had been able to do, of course, for he had better instruments.

It turned out that the stars in one portion of the sky, in the constellations of Lyra and Hercules, seemed, on the whole, to be moving away from each other, as though a kind of ragged hole were opening up in that direction. On the opposite side of the sky, the stars, on the whole, seemed to be moving closer to each other, as though they were shutting a ragged hole.

This could be most easily explained by supposing that

25

the Sun is moving in the Lyra-Hercules direction. In that case, the stars it approaches would be spreading outward, and those from which it is receding would be closing together.

Herschel showed, therefore, that the Sun itself is moving just as the other stars are and that it is not the immovable center of the Universe, any more than the Earth is.

A century and a quarter after Herschel's time, the Dutch astronomer Jacobus Cornelis Kapteyn (1851–1922) undertook to repeat Herschel's work. By Kapteyn's time, telescopes had further improved and, even more important, photography had been invented. Kapteyn could take photographs of randomly selected portions of the sky, and then, at leisure, he could count the stars contained in the chosen squares.

In 1906, he confirmed Herschel's notions of the lens-shaped Galaxy, but found it to be considerably larger than Herschel had thought. By 1920, he estimated the Galaxy to be 55,000 light-years wide and 11,000 light-years thick.

However, the Milky Way remained equally bright all the way around the sky so that Kapteyn, like Herschel, felt the Sun to be at or near the center of the Galaxy.

Kapteyn also repeated Herschel's work on the determination of the proper motion of the stars, to see what regularities he could find in it. Naturally, he could do much more refined work than Herschel had been able to do, and in 1904, he decided that the stars move in two large streams. One stream moves in a particular direction, the other in a directly opposite direction.

Kapteyn's last student was the Dutch astronomer Jan Hendrik Oort (b. 1900), and he considered those two star streams. It seemed to Oort that the Galaxy, like other astronomical objects, ought to be turning about its center. Just as the various planets revolve about the Sun, why should not the Sun (and all the other stars) turn about the center of the Galaxy?

If so, the stars (like the Solar system's planets) might

well all be turning about the Galaxy's center in the same direction. Nevertheless, those stars that are nearer the center ought to be moving more quickly than those stars farther from the center (just as planets near the Sun move more quickly than planets far from the Sun).

By Oort's time, the Herschel-Kapteyn supposition that the Sun is at or near the center of the Galaxy had been shown to be incorrect. The Sun, as was increasingly understood, is a considerable distance from the center of the Galaxy.

In that case, the Sun is circling the center of the Galaxy at some moderate speed, but the stars that happen to be closer to the center are moving faster than the Sun and all are racing ahead, to a greater or lesser extent. On the other hand, the stars that are farther from the center of the Galaxy are moving more slowly than the Sun, and all are lagging behind to a greater or lesser extent.

That would exactly account for Kapteyn's two streams that seem to be moving in opposite directions.

Oort calculated that the Sun is moving about the Galactic center (which is about 30,000 light-years from us) in a fairly circular orbit at a speed of about 220 kilometers per second, relative to that center. This is about 7.5 times as fast as the Earth moves about the Sun, relative to its center.

Nevertheless, so mighty is the Sun's orbit about the Galactic center that it makes one turn about that center in 230 millions years. That is a very long time, but not too long when compared with the duration, so far, of the Solar system. Since the Solar system was formed, it has traveled about the Galactic center nineteen times and is now making its twentieth circuit, assuming that its orbit and its orbital velocity have not changed significantly in all that time.

Probably they haven't, for the stars are so far apart and move so slowly in comparison to those enormous distances, that even if all the stars were moving randomly, the chances that our Sun, in our region of the

Galaxy, would move close enough to another star to have its orbit significantly altered is very small. Since the stars are *not* moving randomly, but are all sweeping along in more or less the same direction, the chances of gravitational interference become even tinier.

This is a good thing, for if the Sun's orbit were altered in such a way as to make it distinctly elliptical, it might penetrate the inner regions of the Galaxy once every rotation (as a comet penetrates the inner Solar system once every rotation), and it would then find itself in a region of dangerous disturbances and radiation. The fact that we are all still here and that life has survived on this planet for well over 3 billion years is itself testimony to the regular and comparatively undisturbed way in which the Sun has been circling the Galactic center so far.

2
A Change of Air

A couple of weeks ago, as I write this, I was at M.I.T., where six people were receiving an award. Each of the six was being introduced by someone else equally important. The group of twelve included three Nobel laureates, of whom the most distinguished was, perhaps, Linus Pauling, a two-time Nobel laureate (Chemistry and Peace), who was eighty-seven years old. He had come to introduce his old teacher, Herman F. Mark, who was ninety-four, and who was one of those receiving an award. It was such a pleasure to see the two elderly fellows beaming at each other. I tried to imagine the feelings of Herman Mark at being introduced by the kid after all those years.

I was one of the six receiving an award, by the way, and gave the feature speech the next night. I didn't feel that I was in a class with the other awardees, but I upheld the honor of science fiction and accepted the quite beautiful award they were handing out.

Before the award ceremonies we had had dinner at

the Boston Museum of Science and from there we were taken to an M.I.T. auditorium by a fleet of limousines. It was a dark and drizzly night, and it was Boston/Cambridge, an area specifically designed for confusion.

So, as was inevitable, our limousine got lost. It tried both sides of the Charles River and made several forays up Massachusetts Avenue. All seemed hopeless. My dear wife, Janet, who has a touching faith in people, pointed out the car window and said, "Ask that man."

We did just that, over my protest that no bystander was ever helpful in such matters. Nor was he, even though he was clearly an M.I.T. student. We then proceeded to ask about seventy-two others. Not one knew where the auditorium was. Most, in fact, seemed to be baffled by the fact that we were speaking English.

Eventually, we encountered a policeman who used his walky-talky and, in the fullness of time, a police car came, collected us, and took us to the right place. We arrived twenty minutes late and Janet's cheerful suggestion that we enter the hall handcuffed, to make a more striking impression as we arrived, was not taken up.

But you know, it is easy to get lost in science, too, especially if you start off on the wrong tack.

In 1798, for instance, a French astronomer and mathematician, Pierre Simon de Laplace, advanced the "nebular hypothesis." He suggested that the Solar system originated from a huge cloud of dust and gas (a nebula), which was slowly spinning and condensing under its own gravitational pull. As it condensed, its rotation speeded up by the law of conservation of angular momentum.

Eventually, the rotation became fast enough for an equatorial bulge to detach itself and eventually condense into a planet. Later, another bulge did the same, and still later another one, and so on.

This hypothesis was a shrewd one, though inadequate in detail, and was immensely popular through the 1800s.

It gave rise to the notion that the farther out from the Sun a planet is, the older it is. Mars, therefore, must be millions of years older than Earth, which is, in turn, millions of years older than Venus.

That made it easy to believe that Mars is inhabited by a race of intelligent beings far beyond us in brainpower and ability, since they have had a so much longer time to evolve. Venus, on the other hand, was thought to be a young world still back in the equivalent of the Mesozoic era—a world of swamps and jungles and dinosaurs and other dramatic life-forms of the past.

Observations of Mars and Venus seemed to back up this view. Mars had ice caps, so it must have water, but from its ruddy color one would naturally suspect that it was mostly desert. Considering its small size and weak gravity, it might have lost much of its water over the eons. For that reason, when the notion of canals arose in 1877, it became easy to imagine a race of superintellects fighting to bring water from the ice caps down to the Martian desert and to plan the invasion and takeover of well-watered Earth.

Venus, on the other hand, had a thick and permanent cloud layer that seemed to indicate that it must be a very watery world. Some people even thought it might be covered by a planetary ocean with no land at all. That was, in fact, the picture I drew of Venus in my novel *Lucky Starr and the Oceans of Venus* (Doubleday, 1954).

By that time, Laplace's original notion had been dismissed long before as totally inadequate. A much subtler and more useful version of the nebular hypothesis had been advanced in 1944 by the German astronomer Carl Friedrich von Weizsacker, and in that one all the planets were formed at about the same time. Mars, Earth, and Venus, we are now quite sure, are all equally old, and there is no longer any reason to think of an ancient Mars and a youthful Venus.

But science fiction writers continued to do so. Old habits are hard to break and, besides, an advanced and

31

malignant race of Martians, and a primitive and dino-saur-ridden Venus, were too dramatic to abandon.

This was encouraged by the fact that even in the mid-1950s, we were almost entirely ignorant of the details of planetary characteristics. In the March 1957 issue of *Astounding Science Fiction*, I had an essay entitled "Planets Have an Air About Them," in which I said a great many perfectly accurate things about gases and planetary grav-itations and types of atmosphere, but was careful not to say a single word about Venus, concerning whose atmosphere we still knew nothing.

A year after that essay, however, everything changed, and here's how it happened.

All objects except those at absolute zero (and there are no objects at absolute zero) give off electromagnetic radiation, if surrounded by an environment at a lower temperature than itself. The higher the temperature, the shorter the wavelength of this radiation. By the time a temperature of about 600° C is reached, some of the emitted radiation is short enough in wavelength to be red light and the object is then said to be "red-hot."

If the temperature continues to rise, shorter and shorter wavelengths of light appear. The object grows orange-hot, yellow-hot, white-hot, and blue-white-hot. If it were hot enough, most of its radiation would appear as invisible ultraviolet.

By the distribution of wavelengths in sunlight, and by the nature of the dark lines, which tell us to what extent different atoms may be ionized, we can tell just how hot the Sun's surface is. We can also tell the surface temperature of any star whose spectrum we can study.

But what about objects that are not quite hot enough to give off appreciable light? In that case, they give off infrared radiation, which has wavelengths longer than that of red light. Infrared does not affect our retinas, so we can't see it, but it is absorbed by our skin and our heat-sensitive nerves can detect it. Thus, if you place your hand near a hot vessel on the stove, you can feel the warmth before you touch it.

If an object cools down still further, it radiates longer and longer wavelengths, until you can't sense them in any way, but the radiation is still there. Beyond the infrared are the still longer wavelengths of microwaves, which are given off copiously by objects that are quite cool to the touch. If we could detect microwaves from a faraway object, then, we could tell from its quantity and wavelength range how hot that object might be.

After World War II (thanks to the development of the technology of radar, which makes use of microwaves), astronomers could build large "radio telescopes" capable of detecting and concentrating small quantities of microwave radiation, just as ordinary telescopes detect and concentrate small quantities of light.

In 1958, a group of American astronomers under Cornell H. Mayer made use of a radio telescope that was sufficiently delicate to detect microwave radiation given off by the dark side of Venus.

How much radiation were they expecting?

That depended in part on how quickly Venus rotates, but in 1958 no one knew anything about Venus's rotation period. One could not see features in the clouds that could be followed as they moved about the planet, and the solid surface beneath the clouds was utterly invisible.

Some astronomers thought that Venus keeps one side permanently to the Sun. If that were so, then the dark side would be in permanent darkness and it would be, at best, quite cold. To be sure, wind from the sunlit side would carry some warmth into the dark side, but perhaps not much (witness Antarctica on Earth during its months of winter). As a result, the microwave radiation from the dark side, at least, might be very small.

On the other hand, some astronomers thought that Venus might have a rotation period near the 24-hour mark, as Earth and Mars do. In that case, the microwave emission might be quite copious even from the dark side, since that would have been exposed to the Sun a few hours before. In that case, microwave emission might indicate a roughly Earthlike temperature, since the fact that Venus is closer to the Sun might be balanced by the fact that its cloud layer reflects most of the sunlight it receives.

Well, Mayer did detect the microwave radiation from Venus and he did not get either expected alternative. He did not get a very low temperature of a dark side that never sees the Sun, nor did he get an Earthlike temperature, nor, for that matter, anything between.

Instead, he got a flood of microwave radiation that indicated a temperature of at least 300° C, some two hundred degrees above the boiling point of water. It was a thunderbolt. No one had expected such a hot Venus.

But *why* should Venus be so hot? Surely the cloud layer should have cooled it off. Besides, Mercury is closer to the Sun than Venus is, and has no clouds to reflect light, or any atmosphere at all, and Venus seems to be hotter than Mercury!

Perhaps, then, Venus's atmosphere, instead of tending to cool the planet by its insulating clouds, actually acts to *warm* the planet.

After all, solar radiation reaches a planetary surface in the form of wavelengths short enough to constitute light. This is absorbed by the planet and the surface is warmed. At night, the warmed surface radiates into the emptiness of space, but the planet, unlike the Sun, is not hot enough to radiate light. It radiates infrared.

Oxygen, nitrogen, and argon, which make up almost all of Earth's atmosphere, are transparent to both light and infrared. As far as they are concerned, light passes through to Earth's surface by day, and infrared leaves Earth's surface by night, with no interference in either case, and a certain equilibrium temperature is maintained.

Carbon dioxide and water vapor, however, are transparent to light but not entirely to infrared. This means that while light has no trouble reaching Earth's surface by day, the infrared radiated by the surface at night is somewhat blocked. The average temperature therefore rises a bit. It is the small quantity of carbon dioxide and water vapor in Earth's atmosphere that makes Earth's temperature milder and, on the whole, more suitable for life.

This is called the "greenhouse effect," because it has often been compared with a greenhouse, in which the glass allows light to enter, but prevents infrared from emerging so that the temperature inside the greenhouse is warm even in the winter. (Actually, many people point out that this is not because the glass blocks infrared, but because it keeps the warmed air itself from escaping, so that it prevents convection rather than radiation. However, there is no hope of changing the phrase.)

Suppose, then, that Venus does not have the kind of atmosphere we thought. As long as we had the notion of a Mesozoic Venus fixed in our minds, we had to assume an atmosphere basically like the Earth's. But suppose that was not so.

Suppose that Venus had a different atmosphere—a change of air. Suppose Venus's atmosphere was rich in carbon dioxide as well as in water vapor. There might have been enough greenhouse effect to raise the temperature of its ocean appreciably higher so that still more water vapor moved into the air. The greenhouse effect would be heightened and the temperature would go up still further, so that carbon dioxide would bake out of limestone, and the temperature would rise still higher. Eventually, the oceans would boil until, finally, Venus would be extremely hot and utterly, completely dry. It would be the result of a "runaway greenhouse effect."

This point of view was vigorously supported by, among others, Carl Sagan and James Pollack.

There were some astronomers, however, who could not let go the picture of a watery Venus. They argued that the rich emission of microwaves might not be the result of surface heat at all, but is merely the effect of electrical phenomena in the upper atmosphere of Venus. It had recently been discovered that Jupiter has a powerful magnetic field and produces microwave radiation that are *not* the result of surface heat. Why not Venus as well?

Was there anything about the microwave radiation that could be used to distinguish between these two possibilities?

For one thing, the microwave radiation was particularly high in the higher wavelengths of 3 centimeters and more. It dropped rapidly at wavelengths under 3 centimeters. Why?

Sagan's explanation was this: If the microwaves originated as a result of Venus's very hot surface, those microwaves would have to pass through Venus's atmosphere in order to get into space and travel to Earth's detecting instruments. Venus's atmosphere might absorb the short-wavelength microwaves and let the long-wavelength microwaves pass through.

If, on the other hand, the microwaves originated high in the atmosphere, they would move into space without passing through significant amounts of matter, so some reason other than atmospheric absorption would have to be found to account for the fall-off at short wavelengths. No good reason suggested itself.

Of course, if astronomers went with the atmospheric absorption suggestion, that created some difficulty. To do all that absorbing, Venus's atmosphere would have to be about a hundred times as dense as Earth's is. But that might prove to be so.

An even better way of distinguishing between the two views existed. Consider the microwaves emerging from the center of Venus's disk. They would travel straight up through the atmosphere to reach space and speed on their way to Earth. Imagine, however, microwaves released near the edge, or "limb," of Venus's disk. To reach Earth, they would have to pass through the atmosphere at a slant and, therefore, through a much thicker layer of gas. There would be more absorption, so that less microwaves would get through.

As a result, the amount of microwaves absorbed would increase steadily as one traveled from the center of Venus's disk toward the limb in any direction. There would therefore be less microwave emission and this is called "radio limb-darkening." (There is a limb-darkening effect on the Sun because its own atmosphere absorbs some of the light it emits, so that this is a well-known phenomenon.)

But suppose the microwaves originated in Venus's upper atmosphere, in the planet's ionosphere, if it had one. There would be no absorption to speak of, either from the center or from the limbs, because there would be little in the way of gases above the ionosphere. However, as seen from Earth, the ionosphere would be thicker at the limbs than at the center, because at the limbs, it would be seen at a slant. As a result, we would detect somewhat *more* radio waves from the limb than from the center. There would be an example of "radio limb-brightening."

In short, if the limb were dimmer than the center, it would mean a hot surface; if it were brighter, it would mean a hot ionosphere and a possibly cool surface. From Earth, however, Venus was so nearly a mere dot of light that astronomers couldn't tell what microwaves came from the center and what from the limb. (Nowadays, a quarter-century later, we have instruments advanced enough to do the job.)

But on August 27, 1962, the United States launched a Venus probe, *Mariner 2*, designed to pass near Venus and to make a variety of measurements while doing so.

On December 14, 1962, *Mariner 2* skimmed by Venus, making its closest approach at 34,831 kilometers (21,648 miles) above the cloud layer. At that distance, Venus's disk was about thirty-five times as wide as that of the Moon seen from Earth.

Mariner 2 measured the intensity of microwaves at a wavelength of 1.9 centimeters across the disk of Venus. The results were unmistakable. There was radio limb-darkening. That was strong support for the suggestion that Venus's surface is very hot.

In addition, *Mariner 2* detected no magnetic field at all. Since a magnetic field is more or less necessary for ionospheric microwave activity, that further weakened the suggestion that microwave radiation is a phenomenon of Venus's upper atmosphere.

Finally, *Mariner 2* made more accurate determinations of the intensity of microwave radiation from Venus than could be done from Earth, and it turned out that Venus is even hotter than had been thought. The surface is not at 300° C, but at 400° C.

Later on, further, more sophisticated probes passed by Venus, and the Soviet Union, in a series of attempts, actually dropped entry capsules into Venus's atmosphere.

By the end of the 1960s, it was clear that Venus's temperature is not 400° C, but more like 480° C. In addition, the atmosphere is indeed as thick as the microwave-absorption suggestion indicated. It is about a hundred times as dense as Earth's atmosphere. Furthermore, in line with the notion of the runaway greenhouse effect, the atmosphere is about 95 percent carbon dioxide while the rest is nitrogen.

(Considering the density of Venus's atmosphere, the total quantity of nitrogen it contains is perhaps five times that contained in Earth's atmosphere, but the quantity of carbon dioxide overwhelms it and makes it a minor constituent, in comparison.)

All this is horrible enough, but what about Venus's

clouds? Ever since the discovery of its cloud layer, astronomers had assumed that those clouds consist of water, as they are on Earth. It might still be so on Venus, as the developing heat of the runaway greenhouse effect forced all the surface water into the upper atmosphere as permanent clouds and, beyond that, into space.

Beginning in 1973, however, astronomers suggested that spectroscopic data made it seem that the clouds of Venus are not exactly pure water, but were, instead, a rather concentrated solution of sulfuric acid. In the late 1970s, Soviet probes reaching into Venus's atmosphere supported this conclusion and showed that there is more sulfur dioxide in Venus's atmosphere than there is water vapor. The sulfur dioxide adds to the greenhouse effect.

So there we have it. Venus has enormous gas pressures, enormous temperatures, a totally unbreathable atmosphere, and clouds of sulfuric acid on high. Sagan commented that Venus is very close to what people have imagined Hell to be like.

In one respect, Venus proves a little better than some have thought. The thick clouds might, after all, block so much light that Venus's surface is shrouded in eternal night. In fact, when the Soviets first sent objects down to Venus's surface, they included floodlights to make it possible to take photographs.

However, about 2.5 percent of the sunlight falling on Venus penetrates the clouds and reaches Venus's surface, making it possible to take photographs without artificial aid. In fact, since the sunlight reaching Venus is twice as intense as that reaching us, it means that Venus is lit one twentieth as brightly as Earth is on a sunny day. That's over a billion times as bright as the full Moon so at least Venus is a lighted Hell and not a shrouded one.

Here's something else. Why doesn't Venus have a magnetic field?

Earth has a diameter of 12,756 kilometers (7,926

miles), while Venus has a diameter of 12,140 kilometers (7,544 miles). Earth has an overall density of 5.5 times that of water, while Venus has one of 5.2 times that of water.

The similarity in size and density makes it seem certain that if Earth has a liquid iron core (as seems to be the case), then Venus must have one, too. (Of the other three terrestrial bodies, by the way, the density of Mercury is 5.4; that of Mars is 4.0; and that of the Moon is 3.3. It follows that Mercury should also have a liquid iron core, while Mars and the Moon do not.)

The current feeling is that Earth has a magnetic field because swirls are set up in the electricity-conducting liquid iron core as a result of the planet's relatively rapid rotation. It follows that the Moon and Mars, without a liquid iron core, should not have a magnetic field, and probes have shown that indeed they do not.

Mercury has a liquid iron core, but its period of rotation is long—1,407 hours, compared to Earth's 24. Apparently, this is enough, however, to allow a very weak magnetic field to be developed.

That leaves Venus. Whether Venus has a magnetic field as strong as Earth's—or stronger—or weaker, depends on its period of rotation, and as I said earlier in the essay, until the 1960s no one had any notion as to how quickly it rotated.

Guesses had been anywhere between 24 hours and, if it rotated on its axis in the time it took it to revolve about the Sun, 5,400 hours.

But suppose a microwave beam is sent to Venus. It would pass through the cloud layer as though it were not there and it would be reflected by Venus's solid surface. If the surface of Venus is motionless, the incoming beam would be unaffected; the reflected beam would return with the same wavelength it had left with. If, however, the surface of Venus is moving, as it would have to if it rotated on its axis, the wavelength of the beam would be altered a bit and this alteration would be detected in the reflected beam. The greater the speed of movement of the planetary surface, the greater the change in wavelength.

On May 10, 1961, a microwave beam was sent out to Venus, and the results were absolutely astonishing. They were announced in 1962 by Roland L. Carpenter and Richard M. Goldstein, even as *Mariner 2* was on its way to Venus.

Venus is rotating on its axis *even more slowly* than it revolves about the Sun. It is unique in the Solar system in this respect, as far as we know.

Venus rotates on its axis in 243 Earth days, or 5,832 hours. We can compare it to other worlds if we put it in terms of the speed of motion of a point on the planetary equator as the planet rotates.

A point on Earth's equator moves at 1,037 miles per hour. A point on Mercury's equator move at 6.7 miles per hour. A point on Venus's equator moves at 4.0 miles per hour. To put it dramatically, Earth is moving like a jet plane; Mercury is moving at a run; and Venus is moving at a walk.

If Mercury has only a very small magnetic field, it wouldn't be at all surprising if the still slower Venus would have so small a tendency to set up swirls in its liquid iron core that its magnetic field would be indetectable.

And as I said, when *Mariner 2* reached Venus, it could detect no magnetic field and this, in itself, served to support the observation that Venus rotates on its axis very slowly.

In parting, let me mention that, just to make Venus even more peculiar, the planet is rotating, at this slow rate, in the "wrong" direction. Instead of turning west to east as the Sun, Mercury, Earth, Moon, Mars, Jupiter, Saturn, and Neptune do, it turns east to west.

Why this is so, we simply don't know—but then we

don't want to solve everything, do we? After having learned so much about Venus in the last thirty years (we have even mapped its surface by beams of microwaves), surely we want to leave something to keep future astronomers busy and happy.

3
The Changing Distance

I keep a professional eye on manifestations of scientific innocence. After all, I am a science writer and a professional "explainer" and I find that the public, if I listen, will tell me what it is that they want explained.

For instance, on September 28, 1988, the planet Mars made one of its closest approaches to Earth. The media, inevitably, made a big thing out of it. "Mars in close approach to Earth," they trumpeted.

I could well imagine that people who knew nothing about astronomy, reading this, would imagine that Mars was peeping over their shoulders, and that its presence (perhaps only a few yards away) had some sort of eerie significance.

I might well have let that go were it not for the fact that I found myself involved in the mystique. I received a phone call from a very pleasant fellow at one of the television networks. He wanted to interview me on the subject of Mars.

The studio was not far off, so my hatred of travel was

not activated, and I agreed to his request. I walked over and took my seat under the lights.

I was sure of the first question, and I was ready.

"How about this close approach of Mars to Earth," asked the interviewer in an awed tone. "What does it mean?"

"Not a darned thing," I replied cheerfully, and explained.

It isn't easy to explain in two sentences but that didn't bother me overmuch because I knew that the particular soapbox of my science essay series was awaiting me. So sit back and let me talk about the distance of some astronomical objects from Earth, and how those distances change.

Let's begin with the Moon, which is, of all sizable bodies, the nearest to the Earth. It goes around the Earth in 27.32 days (relative to the stars), and in doing so, it stays at roughly the same distance from us.

The average distance of the Moon from the Earth, center to center, was calculated with reasonable accuracy even by the ancient Greeks. In the last few decades, however, we've bounced microwaves off the Moon, and from the time it took the microwave beam to go and return, we've managed to determine the distance of the Moon to within a few hundred meters.

The Moon is, on the average, 384,400.5 kilometers (238,906.5 miles) from the Earth.

If the Moon were going about the Earth in a perfect circle, that would be its distance from the Earth at all times, but the Moon's orbit is *not* a perfect circle, but an ellipse. The ellipse is quite close to a circle; so close that if we were to draw the Moon's orbit to scale on a sheet of paper, it would look like a circle to us—nevertheless it is not quite a circle.

Where a circle has one center, an ellipse has two "foci," one on either side of the center. In the case of the Moon's elliptical orbit, Earth is located at one of the foci; that is, to one side of the center of the ellipse.

The straight line that passes through the two foci from one side of the ellipse to the other is the "major axis." When the Moon is at that end of the major axis that is at the same side as is the Earth-occupied focus, it is as close to Earth as it can get. It is then at "perigee," from Greek words meaning "near by Earth." When the Moon is at the other end of the major axis, it is as far from Earth as it can get and it is then at "apogee," from Greek words meaning "away from the Earth." (See Figure 1.)

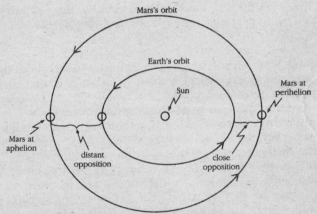

Figure 1. Ellipse (exaggerated)

The Moon, as it travels in its orbit about the Earth, moves from perigee to apogee, and back to perigee. The difference in distance that results isn't much, because the orbit isn't very elliptical. Still, at perigee, the Moon is only 356,375 kilometers (221,451 miles) from Earth, while at apogee, it is 406,720 kilometers (252,736 miles) from Earth.

This difference in distance over the course of the four-week orbit is thus 50,345 kilometers (31,284 miles), or about 13 percent of its average distance.

Does that difference in distance affect the Moon's appearance in any way? It certainly does. The apparent diameter of the Moon when it is at perigee is 33.48 minutes of arc, but at apogee it is only 29.37 minutes of arc. The Moon at perigee is 14 percent wider in appearance than it is at apogee. Its apparent area at perigee is 30 percent greater than at apogee, and this means that if the Moon happens to be full at perigee, it is 30 percent brighter than it is when it happens to be full at apogee.

Would you think that that would matter? Apparently, it doesn't to the general public. They have never seemed aware (as far as I know) of the fact that the full Moon can be 30 percent brighter at some times than at others.

Next, let's consider the Sun. The Earth travels about the Sun once every 365.2422 days. Its orbit, as it does so, is nearly circular so that the Sun remains more or less at the same distance from Earth at all times. Its average distance from the Earth is just about 149.6 million kilometers (93.0 million miles), center to center, or 389 times the distance of the Moon.

But the Earth's orbit is not a perfect circle either; it is also slightly elliptical. Earth's orbit is less elliptical than the Moon's is. The extent of the ellipticity of an orbit is indicated by its "eccentricity." The eccentricity of a circle is exactly zero, but the eccentricity of the Moon's elliptical orbit is 0.055 which, as you see, is not far removed from zero. The eccentricity of the Earth's elliptical orbit is, however, only 0.0167.

Nevertheless, even the slight eccentricity of Earth's orbit means that Earth's distance from the Sun changes measurably in the course of a year. When the Earth is at "perihelion" ("near by the Sun"), the Sun is 147.1 million kilometers (91.4 million miles) from the Earth. At "aphelion" ("away from the Sun"), the Sun is 152.1 million kilometers (94.5 million miles) from the Earth.

The difference in distance is 5.0 million kilometers (3.1 million miles), which is only 5.4 percent of its average distance. The smaller percentage difference in the changing distance of the Sun as compared to the changing distance of the Moon is the result of the fact that Earth's orbit is less eccentric than the Moon's orbit.

The changing distance of the Sun reflects itself in the Sun's apparent size in the sky. When the Earth is at perihelion, the Sun has an apparent diameter of 32.60 minutes of arc; when the Earth is at perihelion, the Sun is only 31.63 minutes of arc in diameter. At perihelion, then, the Sun is 3 percent wider in appearance, which makes it 6 percent larger in apparent area which, in turn, means that it yields 6 percent more light and heat at perihelion than at aphelion.

However, if the 30 percent additional light of the full Moon at perigee goes unnoticed, you can be sure that the 6 percent additional light of the Sun at perihelion is ignored. (This is especially so since perihelion, at this epoch of time, happens to come when it is winter in the northern hemisphere and it is in the northern hemisphere that most human beings live. The additional brightness of the Sun is thus masked by the fact that the Sun is, at that time, lower in the sky and remains above the horizon for less time.)

Even so, in recent decades this perihelion-aphelion difference, combined with the precession of the equinoxes and slight periodic changes in Earth's orbital eccentricity and axial tipping, has been advanced as a cause for long-term swings in Earth's climate, including the production of ice ages. That, however, need not concern us here.

Now we can go on to Mars, which is quite another proposition. Since the Moon circles the Earth, and the Earth circles the Sun, both bodies seem to make a more or less smooth circuit of the sky, traveling at a nearly steady pace from west to east, against the background of the stars (if you discount the effect of Earth's rotation).

Mars, however, circles the Sun as Earth does, but at a different distance and at a different speed, so that you have two separate orbits instead of one. This means that Mars's apparent motion in the sky is far more complicated than that of either the Moon or the Sun.

Whereas Earth is at an average distance of 149.6 million kilometers (93.0 million miles) from the Sun, Mars is at an average distance of 228 million kilometers (142 million miles) from the Sun. This means that Mars is 1.52 times as far from the Sun as the Earth is and, in moving around its orbit, Mars must travel a distance that is 1.52 times that which Earth must traverse.

In addition, Earth, being closer to the Sun than Mars is, is more strongly influenced by the Sun's gravitational pull, and whips along more quickly in its shorter orbit than Mars does in its longer one. Whereas the Earth moves about the Sun at an average speed of 29.79 kilometers per second (18.51 miles per second), Mars moves about the Sun at an average speed of only 24.13 kilometers per second (14.99 miles per second). Therefore, it takes Mars longer to complete a circuit about the Sun than you would expect from its longer orbit. Whereas Earth moves about the Sun in 365.2422 days, Mars moves about it in 686.98 Earth days or 1.88 Earth years.

This means that, as Earth and Mars both race about the Sun in the same direction, Earth is forever overtaking Mars, racing ahead, gaining an entire lap, coming up from behind, overtaking it again, and so on, over and over, for all the billions of years that both have existed.

This further means that Mars is closest to Earth at the moment when Earth catches up with Mars and is about to pass it. Both are then on the same side of the Sun, and at the passing point, a line drawn from the Earth to the Sun will, if extended, pass through Mars as well (see

Figure 2), assuming that both planets are moving in the same plane, which they aren't, quite.

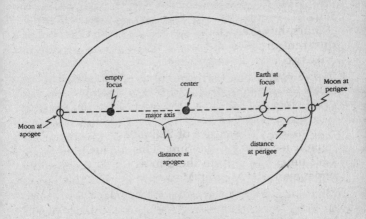

Figure 2. Mars at Opposition and Conjunction

As seen from Earth, it would seem, at that moment of passing Mars, that Mars at midnight would be as close to zenith as it could get, and was exactly on the opposite side of the Earth from the position of the Sun. Because Mars and the Sun are at opposite sides of the Earth, Mars is then said to be in "opposition," and it is at opposition that Mars is closest to Earth.

As Earth moves past opposition, gaining on Mars, it moves farther and farther away from Mars. Eventually, it moves so far ahead that is on the opposite side of the Sun from Mars. At that point, it is as far away from Mars as it can get. Mars, at that moment, is on the other side of the Sun, and as seen from Earth, Mars would seem to have come very near the Sun in the sky. Mars and the Sun are in "conjunction" (from Latin words meaning "to join together"). At conjunction, Earth and Mars are farthest apart.

If Mars were standing still, it would take Earth just half a year to move from opposition to conjunction, and then another half year to move back to opposition. However, Mars is chasing along, too—not as quickly as Earth is, but quickly enough to make Earth take a distinctly longer time to gain a complete lap on Mars.

In fact, it takes Earth, on the average, 779.94 days (2.137 years) to gain a lap and go from one opposition to the next.

Now suppose that Earth and Mars both traveled about the Sun in circular orbits. In that case, at opposition, the distance from Mars to Earth would be Mars's distance from the Sun minus Earth's distance from the Sun. This would come to 78 million kilometers (48.5 million miles).

At conjunction, Earth and Mars would be on opposite sides of the Sun and the distance between them would be 78 million kilometers plus the full width of the Earth's orbit (look at Figure 2, if that isn't obvious to you). Earth and Mars would then be separated by a distance of 377 million kilometers (234 million miles).

The distance at conjunction would then be 4.8 times as great as at opposition, and Mars, at opposition, would shine something like 23 times as brightly as it would at conjunction.

Surely, *that* would be noticed.

Well, yes and no. At opposition, Mars is high in the sky and is visible all night long. As it passes opposition, however, it moves closer and closer to the Sun, and is visible for less and less of the night. Eventually, it is lost in the Sun and is not visible in the night sky at all, or only briefly in the twilight or dawn.

Mars therefore grows less conspicuous, not only because it grows dimmer, but also because it slowly leaves the night sky. To the nonastronomer, these two reasons may be confused.

However, the orbits of Earth and Mars are *not* circular. Earth's is nearly circular, to be sure, but Mars's orbit departs from that ideal by quite a bit. While Earth's

50

orbital eccentricity is, as I said earlier, 0.0167, the orbital eccentricity of Mars is a comparatively whopping 0.0934, markedly greater even than that of the Moon.

This means that, at its aphelion, Mars is 249 million kilometers (155 million miles) from the Sun, while at its perihelion it is only 207 million kilometers (129 million miles) from the Sun. The difference is 42 million kilometers (26 million miles), which is 18.4 percent of the average distance of Mars from the Sun.

For simplicity's sake, suppose we consider Earth's orbit a circle, which is very nearly is.

Opposition can take place anywhere along Earth's orbit. It can take place when Earth passes Mars at the Martian aphelion, or at the Martian perihelion, or anywhere in between.

If opposition takes place at the Martian aphelion, then the distance between Earth and Mars is 249–149, or 100 million kilometers (62 million miles). If opposition takes place at the Martian perihelion, it is 207–149, or 58 million kilometers (36 million miles). Because Earth's orbit is not exactly circular, that minimum distance can sometimes be as small as 55.5 million kilometers (34.5 million miles).

Now forget about the Sun. Consider only oppositions, when Mars is high in the sky and stays in the sky all night long, and when the Sun is exactly on the other side of the Earth. Mars is considerably closer to Earth at some of these oppositions than at others (see Figure 3).

For that reason, Mars, in opposition at perihelion, is 3.25 times as bright as when it is in opposition at aphelion. And *this* is noticeable.

The noticeability is particularly important because of the matter of color. Venus and Jupiter, the two brightest planets, are white. Mars, when it is in opposition at perihelion, is actually slightly brighter than Jupiter, but is distinctly red in color. It is, in fact, the brightest object in the sky that has a red color.

The redness, we know, is because its soil is rich in iron,

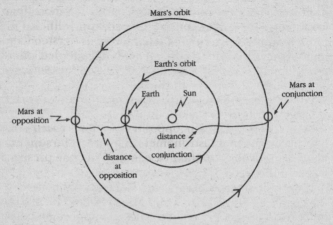

Figure 3. Martian Conjunction at Aphelion and Perihelion

so that what we see is the equivalent of a rusty planet. To the ancients, however, who observed Mars in the days before the iron age, when rust was not a familiar item to them, the color meant only one thing—*blood*.

It is no wonder, then, that the Sumerians, who were the first stargazers in any systematic sense, named the planet in honor of Nergal, their god of war, destruction, and death. The Greeks followed that tradition by naming the planet Ares, after their own god of war, and the Romans called it Mars after theirs. We keep the Roman name.

Naturally, a heavenly object that beams down upon us with the color of blood and with a name that personifies war, destruction, and death is going to be regarded as baleful and threatening. Every other year it shines through the night like a red jewel and every once in a while it shines particularly brightly. At those bright times, when it is in opposition at or near perihelion, it would not be surprising if the hearts of people turned faint and they expected the worst.

That is pure superstition, of course, but it is superstition that rules human hearts and minds far more than cool reason ever does (or did, or perhaps, will).

Even after the astronomical facts were understood and the notion of Mars as a god of war had receded into a merely mythical object, the association of Mars with evil remained.

Prior to 1965, Martian oppositions were viewed with great excitement by astronomers. There was no superstitious fear involved in their case, but rather the presence of great hopes. Once the telescope was invented in 1608, it became possible to view Mars through them and see much more than could be seen by the eye alone. At opposition, when Mars was closest, it could be seen larger and clearer than at other times, and when the opposition came at perihelion, it could be seen largest and clearest of all.

Every thirty-one years or so, Mars is in opposition at or very near perihelion, and then astronomers unlimber their telescopes, prepare for observations, and exude a great deal of excitement that finds its way into the public press. People then marvel over the "close approach of Mars" and perhaps feel a bit nervous about it, too.

At each close approach, of course, telescopes, spectroscopes, photographic techniques, and so on, had made some advance over the situation at the previous close approach a generation earlier. There was therefore always the chance that surface markings might be seen more clearly than ever before, that Mars might be mapped more definitively, that unexpected discoveries might be made, and so on.

In 1877, when Mars made a close approach, the American astronomer Asaph Hall (1829–1907) seized the opportunity to carry on a deliberate search for any small satellites that might happen to be very close to Mars. (They had to be small and close, or they would have been discovered long before.) On August 11, he gave up, but his wife, Angelina Stickney Hall, said, "Give it one more night, Asaph." He did, and discovered two satellites he named Phobos ("fear") and Deimos ("dread") after the sons of Mars.

At that same opposition, the Italian astronomer Giovanni Virginio Schiaparelli (1835–1910) was able to map Mars better than anyone ever had before. He noted narrow dark lines that he took to be waterways and he called them "canali," which was Italian for "channels," a name given to any thin, natural stretch of water like the English Channel.

The word was translated as "canals" in English, however, which was a crucial error, for that word is applied to *artificial* waterways. At once, many nonastronomers (and a few astronomers) felt that evidence had been discovered of life on Mars. And not only life, but an advanced technology capable of constructing canals.

After all, Mars is a small planet with a surface gravity only two fifths that of Earth, so it might be slowly losing its water. The canals would be built in order to bring water from the polar ice caps to the agricultural lands in the warmer part of the planet.

The American astronomer Percival Lowell (1855–1916) took up this view. He established an observatory in Arizona where Mars could be viewed through the thin, dry air of desert uplands. He published maps showing Mars to be criss-crossed with straight canals, meeting at "oases," and wrote a couple of books that proclaimed Mars to be the abode of advanced life.

Most astronomers were skeptical, but the general public ate it up. Not only did they accept the notion of advanced Martian life, but the old superstitions of the bloody planet still exerted their influence and Martian life was assumed to be evil.

In 1898, the British writer Herbert George Wells

(1866–1946) capitalized on this by publishing *The War of the Worlds*, the first book, as far as I know, to deal with interplanetary warfare. He meant it as a social satire. The European nations, Great Britain in particular, had just finished carving up Africa without any regard for the Africans, and Wells pictured the Martians as landing in Great Britain and taking it over without any regard for the British.

The general public, however, ignored the satire and concentrated entirely on the horrors of the Martian invasion and the evil nature of the Martians.

That book fastened onto science fiction what became virtually a convention: Martians are advanced far beyond Earth technology, but they are decadent and evil, and because their planet is dying, they lust to conquer Earth.

I don't know how many Martian invasion stories were written between 1900 and 1965, but virtually every one must have done its bit to bolster the age-long superstitious view of evil Mars—which had arisen only because the poor planet is rusty and is close enough to us so that the rust shows, especially at its near approaches.

Even as late as 1938 (almost exactly fifty years ago, as I write) the dangers of a Martian invasion remained high in human minds. On October 30, 1938, Orson Welles (1915–1985) produced a radio show dramatizing Wells's novel. He changed the site of the Martian landing from Great Britain to New Jersey, and told the tale by means of fictitious news bulletins and government announcements, like those that had characterized the war scare of the previous month, the one that had ended in the West's surrender at Munich to Hitler.

Welles made it quite plain that the story was fiction, but a large number of people in New Jersey went into panic and clogged the highways as they tried to get away from the invading Martians.

I was a little bitter about this in my interview on Mars, when I was asked to comment on the 1938 invasion scare.

I said, "Isn't it sad that you can tell people that the ozone layer is being depleted, that the forests are being cut down, that the deserts are advancing steadily, that the greenhouse effect will raise the sea level 200 feet, that overpopulation is choking us, that pollution is killing us, that nuclear war may destroy us—and they yawn and settle back for a comfortable nap. But tell them that the Martians are landing, and they scream and run."

That statement was edited out and did not appear on the air.

The myth of evil Martians and their advanced technology did not come to an entire end until nearly ninety years after Schiaparelli had unwittingly begun it.

On November 28, 1964, the Mars probe, *Mariner 4*, was launched. On July 14, 1965, it passed within 10,000 kilometers (6,200 miles) of the Martian surface and took a series of twenty photographs that were beamed back to Earth. No canals were shown on those photographs. Only craters, which looked very much like those on the Moon.

Other probes followed and by now Mars has been mapped in detail. On its surface there are not only craters, but extinct volcanoes, an enormous canyon, some jumbled terrains, markings that look for all the world like dried riverbeds, and ice caps that are, in part at least, frozen carbon dioxide. There is also a very thin atmosphere that contains no oxygen to speak of.

What Mars doesn't have is any sign of canals, any sign of liquid water, any sign of life.

So now what does it matter if Mars makes a close approach? No professional astronomer is going to attempt to find out anything about Mars by looking at it through a telescope. The probes have told us far more than any Earth-bound telescope can possibly tell us, and any further information of any importance whatever must come from future probes, with or without people aboard.

To be sure, it is still fun for amateur astronomers to look at Mars at these times of close approach. They will see the Martian orb larger and more clearly than they will at other times.

There's nothing wrong with that, but there's no reason for awe, or for vague foreboding or for calling in people to explain the significance of this supposedly fearsome or eerie phenomenon.

No reason, that is, other than a superstition that is 5,000 years old, and was groundless from the very start.

4
The Moon's Twin

One of the difficulties of learning to read at a very young age, and of beginning at once to read indiscriminately, is that it makes it difficult to look back and say, "This is the occasion on which I first learned about thus-and-so." To me, it seems as though my knowledge of the various subjects on which I write books dates back to the dim mists of my personal prehistory.

As an example, I cannot for the life of me recall whether I grew interested in Greek mythology as a result of my early reading of astronomy, or vice versa. I read both as a little kid and they have grown confused in my mind.

The fault may lie in the fact that the books on astronomy I read in the first decade of my life were very strong on the description of the constellations and on the myths that lay behind them. Long before I had any real idea of what stars were, I learned that Ursa Major, the Great Bear, and Ursa Minor, the Little Bear, had been the nymph, Callisto, and her son, Arcas, before they were placed in the sky. It took me quite a while to gather that constellations are about as important to astronomers as national boundaries are to geologists.

One of the Greek myths involves the nymph Io. (The common English pronunciation, which I always use, is "EYE-oh," but the Greek pronunciation is "EE-oh." Some astronomers have now taken to that.)

Io, the daughter of a river god, had the misfortune to attract the lustful attentions of Zeus. (Zeus, in the Greek myths, was given to making love to every female in sight, to the great annoyance of his jealous wife, Hera.)

Hera caught on (she always did) and, in revenge, changed Io into a white cow and set the monster, Argus, to watch over her. Argus had a hundred eyes, and at any one time, only some were closed in sleep, so that he was an efficient watchman. (Of course, three ordinary people in shifts would do as well, but the Greek gods never thought of simple solutions.)

Zeus thereupon sent Hermes to lull Argus to sleep with a soporific tale, and when all hundred eyes closed and poor Argus was snoring away, Hermes killed him.

Whereupon Hera, not to be outdone, sent a gadfly to sting Io and keep the poor transformed nymph on a perpetual move over the lands of the eastern Mediterranean. Every time the myth describes her as crossing a strait in her wanderings from one land to the other, the strait was named "Bosporus" (meaning "cow-ford").

First she traveled to the western shore of Greece, but the sea between it and Italy was too wide for her to cross. (It is still called the "Ionian Sea" in her honor. She then went north and east, traversing the northern shore of the Black Sea and south to the Caucasus, from where she crossed into the Crimean peninsula by way of the "Cimmerian Bosporus" (now known as Kerch Strait). Then she moved into Thrace and crossed the narrow strait into Asia Minor. This is the "Thracian Bosporus" and it is the only ancient Bosporus that retains its name today. It is the Bosporus on whose shores Istanbul stands.

Then she traveled eastward and southward to India, back to the west through Arabia, and crossed the narrow strait at the southern end of the Red Sea into Ethiopia on the upper Nile. This strait is the "Ethiopian Bosporus," now known as Bab-al-Mandab. Finally, she moved north into Egypt, where she bore Zeus's son and found rest (according to the Greeks) as the Egyptian goddess Isis.

Some think that Io is actually a Moon goddess. The curved horns of the cow represent the crescent Moon. The crescent is set in the sky where the numerous stars (the eyes of Argus) watch over her. All the stars fade when the Sun rises, however, so that the Moon is free to wander over the sky, making a complete circle each month just as Io circled the eastern Mediterranean.

(Keep in mind the thought of Io as a Moon goddess for what follows now, as we turn back to astronomy.)

In January 1610, the Italian scientist Galileo Galilei (1564–1642) discovered four dim starlike objects in the neighborhood of Jupiter, making use of a telescope he had devised—the first to be turned on the sky. As he watched from night to night, it was clear that the four objects circled Jupiter just as the Moon circled Earth. The objects therefore came to be thought of as "the moons of Jupiter."

The German astronomer Johann Kepler (1571–1630) suggested they be called "satellites," from a Latin term for parasites who are hangers-on of rich or powerful men in the hope of gaining occasional crumbs in the way of money or preferment. The term came into use and is much preferable to the word "moon," since that had better be left as the name of Earth's satellite and no other. (Nevertheless, I wrote a book more than thirty years ago named *Lucky Starr and the Moons of Jupiter*, a title that has embarrassed me ever since.)

Naturally, the four satellites needed names and Galileo tried to call them "the Medicean planets" in honor of Cosimo II de Medici, Grand Duke of Tuscany (1590–1621), who was Galileo's patron at the time. Fortunately, that didn't stick and they are called the "Galilean satellites" instead, a much better term.

The German astronomer Simon Marius (1573–1624), who saw the satellites soon after Galileo did, gave them each a mythological name. He named them after four individuals who experienced the embraces of Zeus (whom the Romans identified with their god, Jupiter). He named them, in order of increasing distance from Jupiter, "Io," "Europa," "Ganymede," and "Callisto."

I have told you about Io. Europa was a Phoenician princess whom Zeus, in the form of a bull, carried off to Crete. Callisto was a nymph who bore a child to Zeus and was therefore changed into a bear by the indignant huntress-goddess, Artemis, who insisted that the nymphs serving her remain virgins. When Callisto's son, Arcas, having grown into a man, hunted her down and was about to kill her, Zeus turned him into a bear, too, and placed them both in the sky as constellations.

As for Ganymede, he was a handsome Trojan prince who also suited the eclectic tastes of Zeus. (The ancient Greeks had no objection to bisexuality among their gods.) In the form of an eagle, Zeus carried Ganymede to Olympus to serve him as cup bearer.

Why did Marius pick those particular names rather than those of Zeus's other light-o'-loves? A matter of random choice, I suppose.

Ganymede is the brightest of the Galilean satellites, with a magnitude of 4.5, so Marius, with the casual sexism of the time, gave it the masculine name. (All four Galilean satellites are bright enough to be seen by the unaided eye, but they are drowned out in the glare of nearby Jupiter.) The three female names were distributed, I suspect, randomly.

This brings us to one of those coincidences I love. (As those of you who have been following my essays down through the years know, I ardently collect coincidences in science and history. I attach no mystical importance to them. They are simply coincidences.)

Io, remember, is a nymph whose mythical history can be viewed as an interpretation of the astronomic behavior of the Moon. Isn't it strange, then, that a particular satellite should be named Io at a time when absolutely nothing was known about it but its existence, its brightness, and its motion about Jupiter—and that it should turn out to be an almost exact twin of the Moon in several ways.

To begin with, the Moon has a diameter of 3,470 kilometers (2,160 miles) while Io has a diameter of 3,630 kilometers (2,255 miles). Io's diameter is only 4.67 percent larger than the Moon's. No other body in the Solar system is as close to the Moon in diameter as Io is.

Again, the average density of the Moon is 3,341 grams per cubic centimeter, whereas the average density of Io is 3.55 grams per cubic centimeter. Of all the sizable bodies of the Solar system, Io is the only one with a density so close to that of the Moon. Both bodies, presumably, are made up largely of rocky materials. Neither has a large metallic component (as Earth, Venus, and Mercury do), or a large icy component as other large satellites do.

Of course, since Io is a little larger than the Moon and a little denser, too, Io ends up about 1.2 times as massive as the Moon. That's still fairly close.

Next, let's consider the distance of each satellite from its primary (that is, from the planet it circles). The distance of the Moon from Earth, center to center, is, on the average, 384,401 kilometers (238,867 miles). The distance of Io from Jupiter, center to center, is, on the average, 421,600 kilometers (262,000 miles). That means that Io is 9.7 percent farther from Jupiter than the Moon is from the Earth. That's not extremely close but it is reasonably close.

Only one other satellite imitates the Moon more closely as far as distance from the primary is concerned. Dione, one of the satellites of Saturn, is 377,000 kilometers (234,000 miles) from Saturn, center to center. The difference in distance from the primary between Dione and the Moon is only 2.1 percent. Dione, however, is only about 1120 kilometers (700 miles) in diameter, so that it is a small object compared to Io and the Moon.

But the twinship only goes so far. In respects other than size, density, mass, and distance from the primary, the two twins are not similar at all.

To begin with, they circle planets that are enormously different from each other. Jupiter is far larger than Earth, having a mass 318.4 times that of our puny little planet. That means that, at equal distances from the planet (provided that the distances are greater than the radius of either planet) the intensity of Jupiter's gravitational field is 318.4 times that of Earth's.

Io may be at roughly the same distance from Jupiter that the Moon is from Earth, but under Jupiter's mighty gravitational lash, it travels much more quickly. The Moon lazes along its orbit at an average speed of 1.03 kilometers per second (0.64 miles per second or 2304 miles per hour). Io, on the other hand, to keep from being dragged down to destruction by Jupiter's strong attraction, must move along at a speed of 17.4 kilometers per second (10.8 miles per second, or 38,880 miles per hour). In other words, Io moves through space, relative to Jupiter, 17 times as fast as the Moon does relative to Earth.

Since they are at similar distances from their primaries, each satellite must travel a similar distance in orbit in order to make one complete circle about its primary. The Moon's orbit about Earth is 1,207,630 kilometers (750,421 miles) long. Io's orbit about Jupiter is 1,324,500 kilometers (823,000 miles) long.

The Moon completes one orbit about Earth, relative to the stars, in 27.32 days—which is referred to as the "sidereal month," where "sidereal" is from a Latin word meaning "star." Io, however, completes its slightly longer orbit, relative to the stars, in only 1.77 days. Io whirls about Jupiter 15.4 times while the Moon makes a single circuit about the Earth.

A planet does not pull upon its satellite all in one piece. The near side of the satellite is closer to the planet, and is therefore attracted more strongly than the far side of the satellite. As a result, the satellite is stretched in the direction of its primary. There is one slight bulge toward the primary, and another away from it, and this is the "tidal effect."

The extent of the tidal effect becomes greater as the gravitational pull of the primary increases, as the size of the satellite increases, and as the distance between the two decreases.

The existence of a tidal bulge interferes with the rotation of a satellite. The gravitational pull of the primary drags at that bulge as the satellite rotates, and successive portions of the satellite's surface swell toward the primary and then subside. The internal friction produced by such movements in the body of the satellite bleeds away rotational energy, converting it to heat. The rotation of the satellite slows and eventually stops (relative to its primary) so that the bulge remains pointed permanently toward and away from the primary and there is no further drag.

The Moon produces tidal bulges on the Earth, but the Moon is only $1/81$ as massive as the Earth so that its gravitational pull is quite weak, while the Earth's rotational energy is quite large. Although the Moon's tidal effect has slowed the Earth's rotation and made the day considerably longer than it used to be in eons past, it has not yet managed to stop Earth's rotation altogether, relative to the Moon.

The Earth's tidal effect on the Moon, however, is much stronger than the Moon's on Earth, and the Moon has considerably less rotational energy than the Earth has. As a result, the Moon's rotation has stopped, relative to the Earth, and it faces only one side toward us. However, although the Moon does not rotate relative to the Earth, it does rotate on its axis, relative to the stars. Its sidereal day is 27.32 Earth days long, which is exactly equal to its sidereal month (something that is always true of a satellite that has one side perpetually facing its primary).

Io's tidal effect on supermassive Jupiter is insignificant, but Jupiter's tidal effect on Io is enormous—a couple of hundred times that of Earth upon the Moon. It is not at all surprising, then, that Io, and indeed, the other Galilean satellites as well, face one side to Jupiter at all times. Io's sidereal day therefore equals its sidereal month, and each stands at 1.77 Earth days.

Tidal effects tend to pull a satellite into the equatorial plane of the primary, if it is not there already, and to make its orbit circular, if it is not that already. Unless the tidal effect is tremendous, however, the change is a slow one.

The Moon, for instance, is not in Earth's equatorial plane but is tipped about 23 degrees to it. And the Moon's orbit is distinctly elliptical, with an eccentricity of 0.055. This is not a very great eccentricity, being less than that of the orbits of Mars and Mercury, and being about equal to that of the orbit of Saturn. Still, it is more than three times the eccentricity of Earth's orbit about the Sun, and is greater than the eccentricity of any of the orbits of the other large satellites.

With time, the Moon's orbit may become more circular

65

and may approach Earth's equatorial plane. However, the Moon is so distant from Earth, which is so small compared to the giant outer planets, that the tidal effect is comparatively small. Add to this the complicating factor of the Sun's pull from its comparatively short distance of 149 million kilometers (93 million miles) and we would expect the correction of the Moon's orbit to be a long-drawn-out process indeed.

The far greater tidal effects of Jupiter on its Galilean satellites, and the lesser effect of the more distant Sun, makes the situation different there. The orbits of the Galilean satellites are all very close to Jupiter's equatorial plane and their eccentricities are very close to zero.

The tidal effect of Jupiter increases rapidly as the distance of a satellite decreases. It is thus far stronger on Io than on any other Galilean satellite and its orbit would be expected to be more nearly in the equatorial plane and more nearly circular than would be true in the case of the other three. (In fact, Io suffers a greater tidal effect, thanks to the enormous mass of Jupiter and to its own large size for a nonplanetary object, than any other body in the Solar system.)

Now let's consider the combination of eccentricity and tidal effect. The tidal effect decreases as the cube of increasing distance. Because of the Moon's comparatively large orbital eccentricity, it is as close as 356,000 kilometers (221,000 miles) from Earth's center at its closest approach ("perigee") and is as far away as 407,000 kilometers (253,000 miles) two weeks later at its point of farthest recession ("apogee"). The tidal effect of Earth upon the Moon is therefore 50 percent greater at the Moon's perigee than at its apogee.

We can imagine the Moon's surface slowly straining into a slightly greater bulge, fore and aft, and then collapsing into a slightly smaller bulge and then back to a slightly greater one. It would be a kind of accordion effect with a total period from greatest bulge to greatest bulge of one sidereal month, or 27.32 Earth days.

The Moon's twin sister, Io, has a tidal effect upon it that is much greater than that of Earth on the Moon, but if its orbit were precisely circular (as, from what I have said, you would think it should be) there would be no accordion effect at all.

However, Io's orbit *is* slightly eccentric and will stay so because there is interference from other sizable bodies nearby. The gravitational pull of Europa, Ganymede, and Callisto (particularly Europa, which is nearest) exert perturbing effects on Io's orbit that drives it away from the perfectly circular.

This means that there *is* an accordion effect on Io and, indeed, on all the Galilean satellites. However, the effect is stronger and more rapid, the closer the satellite is to Jupiter. Thus, Callisto accordions once every 16.69 days, while Io does so, much more strongly, once every 1.77 days.

These accordion effects use up rotational energy, and this is converted into heat. In effect, then, the Galilean satellites are heated by the tidal effect, and they are heated more strongly the closer they are to Jupiter. Io is more strongly heated by tidal forces than any other body in the Solar system.

This explains the densities of the satellites as we now know them. Callisto, the outermost Galilean, has an average density of 1.83 grams per cubic centimeters, and must be composed mostly of icy materials. Ganymede, the next one toward Jupiter, is more strongly heated and has lost some of its ice, ending with an average density of 1.93 grams per cubic centimeter. Europa has lost still more ice and has a density of 3.04 grams per cubic centimeter. Io, the most strongly heated, has a density of 3.55 grams and must be composed of rocky material entirely—like our Moon.

So now it begins to appear as though Io, which, to start with, seemed to be our Moon's twin, is a unique body that might well be worth a close look.

The first chance for a close look at Jupiter and its satellites came with a probe, *Pioneer 10*, that was launched on March 2, 1972, and that made its closest approach to Jupiter on December 3, 1973. It was designed to take photographs of Jupiter itself rather than of the satellites, but even so it found out something.

Considering the surface gravity and surface temperature of the Galilean satellites, it was not to be expected that any one of them would have a substantial atmosphere. Io was expected to be as airless as its twin, the Moon. (Bodies of comparable size such as Titan, a satellite of Saturn, and Triton, a satellite of Neptune, together with the small planet, Pluto, have atmospheres that are fairly substantial, but they are all considerably colder than the Galilean satellites and can hold on to the correspondingly more sluggish molecules.)

However, *Pioneer 10* sent radio signals to Earth, and when some of these happened to skim by Io, they were distorted in a way that indicated the presence of an ionosphere in the neighborhood of the satellite—a region rich in charged particles. This, in turn, seemed to indicate the presence of an atmosphere generally, even if only a very thin one.

The companion probe, *Pioneer 11*, launched on April 5, 1973, made its closest approach to Jupiter on December 2, 1974, but added nothing to this picture. Still, astronomers had grown curious and had begun to zero in on Io from their Earth-bound observatories.

In 1973, they found definite signs that there is a thin cloud of sodium vapor about Io spreading out to a distance of hundreds of thousands of kilometers. Later, they found that there is an even more extensive cloud of sulfur and oxygen about Io. This cloud, in fact, fills all of Io's orbit, so that it is a thin doughnut of vapor through which Io plows.

Nothing like this had been found anywhere else in the Solar system and it seemed to match, in strangeness, Io's ferocious accordion effect. It couldn't be a coincidence that the particular world that experiences the strongest tidal distortions should also be the only world to fill its orbit with gas.

Since Io simply can't hold a static atmosphere, it cannot merely be bleeding vapor out of such an atmosphere into its orbit. Even if it had started with an atmosphere, that would long since have bled into the orbit and disappeared. Since the doughnut of vapor still exists today, it must be that Io is producing vapors continuously from its inner structure.

Toward the end of the 1970s, two new Jupiter probes were sent outward, probes that were far more sophisticated than the *Pioneers. Voyager 1* was launched on September 5, 1977, and passed Jupiter on March 5, 1979. *Voyager 2* had been launched two weeks earlier, on August 20, 1977, but did not pass Jupiter till July 9, 1979.

While they were on their way, astronomers were studying the accordion effect, and it seemed to some that Io would be heated so strongly that heated material would burst through its outer crust, forming volcanoes that would spew material into the airless void above Io's surface.

A paper to that effect was published only days before *Voyager 1* approached Jupiter and took the first close-up pictures of Io.

Astronomers had naturally expected that a world the size and density of the Moon would have a surface like that of the Moon. However, as *Voyager 1* passed within 19,000 kilometers (12,000 miles) of Io, the photographs of the surface turned out to be very un-Moonlike indeed. The surface of Io showed few craters and it was a melange of red, orange, and yellow, with a little black and white.

What's more, there were indeed volcanoes—and not just dead ones like those on Mars and (probably) Venus. There were volcanoes that were actively spewing out material. Nine active volcanoes were counted, and Io proved to be the first world in the Solar system, other than Earth itself, to show them. When *Voyager 2* passed, a few months later, eight of the nine were still erupting.

They were erupting sulfur dioxide, which broke up into sulfur and oxygen under the lash of the ultraviolet light from the distant Sun. Some of the sulfur fell as a kind of reddish snow, piling up five or six centimeters per year. It filled in most of the craters, so that only the latest still showed blackly, and it was what gave the surface its color. Some sulfur vapor along with oxygen formed a kind of excessively thin atmosphere over Io—perhaps a billionth as dense as that of Earth—and this leaked slowly out into space to form the doughnut of sulfur and oxygen vapor through which the satellite passed.

Worlds have a habit of having different hemispheres. Earth has one hemisphere that is almost all ocean, while the other is crowded with land. The Moon has all its maria in one hemisphere, none in the other. Iapetus has a dark hemisphere and a light one. Mars has a cratered hemisphere and a noncratered one. In the same way, Io has a hemisphere with large volcanoes and one with small ones. None of the reasons for any of these divisions are known.

Io's volcanoes color not only its own surface but that of neighboring satellites as well. Inside Io's orbit is a tiny satellite, Amalthea. It is only 181,300 kilometers (112,000 miles) from the center of Jupiter, and circles the planet in 12 hours. It is only about 240 kilometers (150 miles) across, and it seems to be colored red. Undoubtedly, it has caught some of Io's sulfur snow. There are traces of sulfur on the surface of Europa, too, which is the satellite next farther out from Io.

We have a chance for another look at Io. A probe

named *Galileo*, the most sophisticated one yet, was launched in late 1989, to take a long roundabout path and to reach Jupiter in 1995. It was originally slated for launching in 1982, but there were delays, not the least of which was occasioned by the *Challenger* disaster.

It will swing by each of the Galilean satellites—several times in the case of the outer ones. Io will get only one pass, however, because it is too close to Jupiter to take too many chances of damaging Galileo's delicate instruments by Jupiter's huge magnetosphere with its dense collection of charged particles.

That one pass will be a lulu, however. It will come within 1,000 kilometers (620 miles) of Io's surface, and it ought to catch some of the volcanoes in beautifully detailed action.

How splendid that would be!

5
Worlds in Order

I was at a small science fiction convention here in Manhattan a month ago, as I write this, and I was accosted by a young writer. The following conversation took place.

WRITER: Dr. Asimov, I've been trying to write for a number of years and I've managed to sell a couple of items.

ASIMOV: Congratulations! I'm happy to hear it. Keep it up.

WRITER: I've used you as my model. I've read a lot about you and I thought that I would try to write the way you do. Easily. Copiously. All that sort of stuff.

ASIMOV (cautiously): And have you managed?

WRITER (frowning): No, I haven't. I have to keep thinking about what I write, and more thinking. And rewriting. And starting over. And getting stuck for periods of time.

ASIMOV (uneasily): I'm sorry.

WRITER (perceptibly angrier): I couldn't figure out what was wrong with me. So I talked to other writers, and I found they all have the same trouble. Just like me. There's nothing wrong with me.

ASIMOV *(relieved):* I'm sure there isn't.

WRITER *(pointing his finger in a controlled fury):* But I'll tell you what. There is something seriously wrong with *you!*

ASIMOV *(wincing):* If you had consulted me in the matter, I'd have told you that to begin with. *(But he turned and stamped away.)*

I've never denied that I was peculiar, you see—and not just in my writing technique, but in many ways. I accept that, know about it in detail, and manage to live with it. After all, we're all peculiar in some way or other, and happy is he (or she) who knows the nature of the peculiarity and can make it work for him (or her).

One of my peculiarities is that I love to count and measure and compare and make lists—orderly lists. Why that should be, I don't know, unless it's just to keep my mind occupied when there is any danger of its lying fallow. (I can't bear having my mind go into neutral, and if necessary, I'll count the holes in acoustical tiles in the ceiling rather than allow it to do so.)

In any case, the recent *Voyager 2* flyby of Neptune has slightly upset the order of the listing of Triton among the objects of the Solar system, so I thought I'd expose all of you to the matter of this particular peculiarity of mine—but please don't get annoyed with me because of that, after the fashion of the young writer at the recent convention.

The Solar system contains countless bodies. A very few are large, but many are mountain-size, many more boulder-size, still more pinhead-size, and dust-size, even down to atomic-size. It is inconceivable that I (or anyone) can list all the bodies of the Solar system in order of size, and I won't try. I'll just list the twenty-seven largest bodies in order—with some discussion along the way, of course.

Obviously, the largest object in the Solar system is the Sun, but the extent to which it predominates is not always grasped. Diagrams of the Solar system show the out-sweep of the planetary orbits and, at the center, a tiny, tiny Sun, and that gives us a false idea.

Actually, the Sun has a mass that is 333,000 times that of the Earth. If you could imagine a huge pair of scales operating under a vast gravitational field, and if you put the Sun in one of the scales, you would have to pile in 333,000 objects the mass of the Earth in the other, in order to balance them. Or if you prefer, if the Sun were a million dollars, the Earth would be three bucks.

The Sun is, on the average, less dense than the Earth and, kilogram for kilogram, takes up more room. The volume of the Sun is 1,303,000 times that of the Earth. Suppose you imagine a huge hollow container the size and the shape of the Sun. Imagine taking up a solid object the size of the Earth, grinding it into dust and pouring that dust into the Sun container. You would have to grind up 1,303,000 Earths to fill the Sun container.

But suppose we consider not only the Earth, but all the planets that circle the Sun; all the satellites that accompany the planets; all the asteroids and meteoroids; all the comets near and far; all the dust. All this material that circles the Sun we can call the "planetary system."

The entire planetary system has 448 times the mass of the Earth alone. That, however, means that the Sun, all by itself, is 743.3 times as massive as all the myriad of bodies that circle it. Another way of putting it is that the Sun contains 99.866 percent of all the mass of the Solar system.

Some dispassionate observer, viewing the Solar system with cold eyes attuned only to the presence of mass, might feel justified in saying that the Solar system consisted of a single luminous Sun, with some inconsiderable dregs of nonluminous matter circling it.

It is only because we live on one of those dregs, and know that it carries a rich load of life, that keeps us from stopping the study of the Solar system with the Sun alone. (We are forced by circumstances to consider all other stellar systems as made up of luminous bodies only. We have no detail on any planetary systems other than our own.)

It is very difficult, of course, to grasp numbers in the hundreds of thousands and in the millions, so let us compare the Sun and the Earth on the basis of diameters. This is a one-dimensional comparison that is the cube root of the three-dimensional comparison of volume, and therefore gives us a much smaller figure, more easily grasped. The diameter of the Sun is 109.25 times that of the Earth. Whereas the Earth's diameter is 12,756 kilometers (7,914 miles), that of the Sun is 1,394,000 kilometers (866,000 miles).

Mass is a more fundamental property of an object than is diameter, of course, and I will mention mass when that is useful; but I will cling to diameter for the purpose of visualization.

Now let us shift to the planetary system and note that its most prominent members are four planets that, by Earthly standards, may be considered giants. They are Jupiter, Saturn, Uranus, and Neptune in that order of distance from the Sun.

Of these, Jupiter is the largest. Just to make sure you understand that "large" is a relative term, the Sun has 1,048 times the mass of Jupiter, and has a diameter 9.8 times that of Jupiter. It would take nearly ten Jupiters side by side to stretch across the full width of the Sun.

If, however, we forget the Sun and confine ourselves to the planetary system only, then Jupiter is an impressive giant indeed. Its mass is 317.83 times that of the Earth and its equatorial diameter is 11.8 times that of Earth. Roughly speaking, Earth is to Jupiter as Jupiter is to the Sun.

We can be a bit more dramatic. Jupiter contains 71 percent of all the mass of the planetary system. It is 2.5 times as massive as all the other planets, satellites, comets, asteroids, meteoroids, and dust of the planetary system put together.

Suppose we add to Jupiter the other three giant planets. Saturn has a mass equal to 95.15 times that of Earth; Uranus has a mass equal to 14.54 times that of Earth; and Neptune has a mass equal to 17.23 times that of Earth.

These three giant planets, taken together, have a mass equal to 126.92 times that of Earth, but even so, these three, taken together, have a mass that is only ⅖ that of Jupiter alone. Now add Jupiter, and all four giant planets have a mass equal to $\frac{1}{750}$ that of the Sun.

The four giant planets, taken together, make up 99.25 percent of the total mass of the planetary system (that is, remember, the Solar system minus the Sun). This means that the Sun and the four giant planets (the five largest objects in the Solar system) make up 99.999 percent of the mass of the Solar system. All the matter in the Solar system outside the Sun and the four giant planets make up not quite $\frac{1}{100,000}$ of the whole.

An observer studying the Solar system dispassionately, and finding himself capable of bringing the giant planets to his notice, could reasonably say that the Solar system consisted of one star, four planets, and some traces of debris.

In one respect, incidentally, the giant planets outmeasure the Sun. The original cloud of dust and gas that formed the Solar system turned slowly on its axis and, therefore, possessed a quantity of what is referred to as "angular momentum." Any system that remains isolated always contains the same quantity of angular momentum, neither losing nor gaining any with time.

The angular momentum depends on both the speed of rotation and the average distance of the rotating parts from the center. If one of these properties decreases, the other must increase, and vice versa. As the original cloud contracted, and the distances of its parts from the center decreased, the speed of rotation increased. All the bodies of the planetary system are now rotating at various speeds about their axes, and are revolving about the Sun, or about a planet which is revolving about the Sun as well. The Sun also rotates about its axis.

The Sun, however, has retained only about 2 percent of the total angular momentum of the Solar system. The other 98 percent is to be found in the planetary system and, chiefly, in the four giant planets. Jupiter, all by itself, contains over 60 percent of the angular momentum of the Solar system. The four giant planets, taken together, contain 97 percent of the angular momentum, while all the bodies of the planetary system other than those four possess the remaining 1 percent of the angular momentum.

This represents a serious problem. How can so much of the angular momentum be concentrated in the relatively small planets and so little in the vast central Sun? This remained a nagging problem that was solved in recent decades, by taking electromagnetic fields into account.

Consider planetary rotations. Jupiter, the largest planet, rotates about its axis in 9.9 hours. Saturn, smaller and more distant from the Sun, has a longer rotational period of 10.6 hours. Uranus, still smaller and more distant, has one of 17.2 hours. Well, then, does this mean that the rotation period grows longer with greater distance, or with smaller size?

The answer might lie with Neptune, which is more distant than Uranus but is also slightly larger. The *Voy-*

ager 2 flyby in August 1989 showed that Neptune has a shorter rotational period than Uranus, rotating in 16.0 hours. Apparently, then, the period decreases with increase in mass, regardless of distance. Perhaps the more massive the planet, the greater the supply of angular momentum it gathered.

Then, too, the atmospheric activity of a planet depends upon the temperature differentials maintained within it, and it would seem that this differential would have to depend on the heat received from the Sun. If we set the heat per unit surface area received from the Sun by Jupiter as 1, then that received by the more distant Saturn is about 0.30, while Uranus receives 0.074 and Neptune 0.030.

It is no surprise, then, that Jupiter has a furiously active atmosphere; that Saturn's is milder; and that Uranus (receiving only $\frac{1}{13}$ the heat of Jupiter) is a bland and relatively quiet planet. But what about Neptune? It gets only $\frac{1}{33}$ the heat that Jupiter does and only 2.5 the heat that Uranus does. Surely, it should be even quieter than Uranus.

Well, it isn't. The *Voyager 2* flyby showed that Neptune's atmosphere has winds of 640 kilometers (400 miles) an hour and more. Neptune also has a "blue spot" very much the shape, and in very much the relative position, of Jupiter's "red spot."

You might argue, of course, that the Sun is not the only possible source of heat for the giant planets. The planetary cores are extremely hot, and heat may leak up to the surface from below. The larger the planet, the greater the central heat, and the greater the contribution thereof to atmospheric activity. Neptune may be considerably farther from the Sun than Uranus is, but it is also more massive than Uranus and might that account for the additional heat required for its active atmosphere? Somehow, I don't think so. Neptune is less than 20 percent more massive than Uranus and surely that isn't

78

enough to account for the difference in activity. It's a mystery.

It's time to get to the diameters of the giant planets. In rapidly rotating objects, the diameter varies with direction, because there is an equatorial bulge due to a centrifugal effect. The "equatorial diameter" is the longest and I will use that.

For Jupiter, the equatorial diameter is 142,800 kilometers (88,700 miles); for Saturn it is 120,660 kilometers (75,000 miles); for Neptune it is 50,200 kilometers (31,200 miles); and for Uranus it is 49,000 kilometers (30,440 miles). These bodies are, therefore, the second, third, fourth, and fifth largest objects in the Solar system.

Let us now move on to objects smaller than Uranus. The one that is in sixth place happens to be (hurrah!) the Earth.

It isn't much of a world compared to the objects we've mentioned, to be sure. Earth's mass is a little less than 7 percent that of fifth-place Uranus.

Look at it in another way, though. I have said that everything in the Solar system other than the Sun and the four giant planets makes up only about $1/100,000$ of the whole, and may be regarded as "traces of debris." If we leave out the comets (whose numbers and total mass we can only make vague guesses concerning), we can say that the Earth's mass is just about half the total of these traces of debris.

Earth is found in the "inner Solar system," which includes those regions closer to the Sun than the orbit of Jupiter. In the inner Solar system there are to be found other objects, of course, and three of them are, by Earthly standards, sizable planets.

One is Venus, which is almost as large as Earth is. Its mass is 81.5 percent that of Earth, or just over $4/5$. In terms of mass, Venus is to Earth very nearly as Uranus is to Neptune, so that just as Uranus and Neptune are the twin planets of the outer Solar system, Earth and Venus are the twin planets of the inner Solar system.

Venus and Earth are twins only in size, however. In every other respect, they are startlingly unlike. Earth is, as we know, moderate in temperature, with a water ocean, and with swarms of life. Venus is extremely hot, utterly dry, and utterly dead. Earth rotates in 24 hours west to east, while Venus rotates in 244 days east to west. Earth has a thin atmosphere loaded with oxygen, while Venus has a thick atmosphere that is almost entirely carbon dioxide; and so on.

Earth and Venus together make up almost $7/8$ of those traces of debris I talk about.

In terms of diameter, Earth's is 12,756 kilometers (7,926 miles) while Venus, the seventh largest object in the Solar system, has one of 12,140 kilometers (7,544 miles).

That brings us to Mars, the eighth largest object in the Solar system. Mars is so famous a world, and we talk so much about it, that most people probably don't realize how small it is. Once you talk about its polar ice caps and its rotation period of 24.6 hours, its canyons, its volcanoes, its dried riverbeds, you begin to think of it as an Earthlike world, and therefore perhaps an Earth-sized world.

However, it is not. The mass of Mars is only about $1/10$ that of Earth and only $1/8$ that of Venus. Its diameter is 6,790 kilometers (4,219 miles), only a little over half that of Earth. Its surface area is only 28.3 percent that of Earth. Mars has no surface water, however, so that all its surface is land area and this is just about as large as Earth's land area. Small as it is, Mars is a respectable world.

Of all the objects smaller than Mars that circle the Sun and that may therefore be called planets in the widest definition of the term, the largest is Mercury, yet it is *not* the ninth largest object in the Solar system.

In addition to the planets, there are other objects that circle one planet or another and that are carried, in the grip of that planet's gravitational field, around the Sun. These planet circlers are "satellites."*

Satellites are, on the whole, much smaller than planets are. Not one satellite, for instance, is the size of Mars. Two of the satellites, however, are larger than Mercury. One of these is Jupiter's largest satellite, Ganymede, which has a diameter of 5,262 kilometers (3,270 miles). Then comes Saturn's largest satellite, Titan, which has a diameter of 5,150 kilometers (3,200 miles). Compare these with the planet Mercury, which has a diameter of 4,878 kilometers (3,031 miles).

By diameter, then, Ganymede is the ninth largest object in the Solar system, Titan is the tenth, and Mercury is the eleventh.

Here, though, we are running counter to mass. Ganymede is 780 million kilometers (485 million miles) from the Sun, and Titan is farther still, at 1,425 million kilometers (885 million miles). Both are cold worlds made up largely of icy materials that are comparatively light. (Titan is even cold enough to retain a rather thick atmosphere.)

Mercury, however, is very close to the Sun, remaining at an average distance of 58 million kilometers (36 million miles) from it. Mercury is therefore a hot planet formed out of materials that can stand the heat—rock and metal.

Mercury's rock and metal are far denser than the ices of Ganymede and Titan. Therefore, though the volume of Mercury is only 85 percent of Titan and only 80 percent of Ganymede, Mercury is 2¼ times as massive as Ganymede, and 2½ times as massive as Titan.

---

*In the press and TV, they are often called "moons," which is incorrect. "Moon" is the name given to Earth's satellite and to call other satellites "moons" is about the equivalent of calling other planets "earths."

81

Does that mean that we should consider Mercury larger than either satellite, and give *it* the ninth place? No, for two reasons. First, mass is harder to measure than diameter for the smaller objects of the Solar system and we couldn't make up a neat listing with mass as the basis. Second, mass isn't visible to the eye, while diameter is. If we were to place scale models of Ganymede, Titan, and Mercury side by side, anyone looking at the three globes would judge Mercury to be the smallest, so that's how we'll leave it.

There are five other satellites that are of respectable size and that can be lumped together, with Ganymede and Titan, as "the large satellites." The five include three satellites of Jupiter: Callisto, Io, and Europa. It also includes Neptune's largest satellite, Triton, and Earth's only satellite, the Moon.

Of these five, Callisto is the largest and is therefore in twelfth place, while Io is next and is in thirteenth place. Callisto has a diameter of 4,800 kilometers (2,980 miles) and Io has one of 3,630 kilometers (2,260 miles).

That brings us to fourteenth place, and until recently, it was thought that Neptune's satellite, Triton, was in that slot. Of course, Triton is so far away that there isn't a hope of being able to measure its diameter with anything like accuracy by Earth sightings. Triton's apparent brightness could be measured, however. If Triton reflected the same percentage of the light that falls upon it as other distant satellites do whose diameter *is* known, then, from Triton's brightness, and allowing for its distance, its diameter could be estimated. Its diameter was therefore thought to be 3,500 kilometers (2,175 miles), which would have placed it in fourteenth place.

The *Voyager 2* flyby of Neptune, however, got a close look at Triton and found that its surface is slicked with frozen methane that reflects most of the feeble light of the distant Sun that falls on it. Triton is therefore shinier than was thought, and a considerably smaller Triton would reflect enough light to make it appear as bright as it does in Earth's telescopes.

The fourteenth place was therefore passed on to the Moon, which has a diameter of 3,475 kilometers (2,160 miles). In fifteenth place is Europa with a diameter of 3,138 kilometers (1,950 miles). And it is in sixteenth place that we now find Triton, with a diameter of 2,735 kilometers (1,700 miles).

Your Earthly patriotism may be disturbed at the fact that there are four satellites larger than our Moon, but remember that the Earth is a small planet to have so large a satellite. Of the seven largest satellites, the Moon is by far the largest relative to the planet it circles.

Ganymede, Jupiter's largest satellite, has a diameter that is only 3.7 percent ($\frac{1}{27}$) that of Jupiter. Titan, Saturn's largest satellite, does a bit better, for Titan's diameter is 4.3 percent ($\frac{1}{23}$) that of Saturn. Triton, Neptune's largest satellite, has a diameter that is 5.4 percent ($\frac{1}{18}$) that of Neptune. The Moon, however, has a diameter of 27 percent (over $\frac{1}{4}$) that of Earth. You can almost think of Earth and Moon as a double planet.

So far, we have considered the Sun, eight planets, and seven satellites, filling up the first sixteen places. Where next?

There is one more planet, with its orbit lying almost entirely beyond that of Neptune, and it is Pluto. It was thought to be quite large when first discovered, but closer and closer inspection has caused it to seem ever smaller.

It turns out that Pluto is smaller than any of the seven large satellites and has a diameter of only 2,500 kilometers (1,550 miles). It is in seventeenth place.

After that, there are four satellites to be considered. There are Uranus's two largest satellites, for instance. Of these, Titania has a diameter of 1,610 kilometers

(1,000 miles) and Oberon has one of 1,550 kilometers (960 miles). Following them are Saturn's second and third largest satellites: Rhea, with a diameter of 1,530 kilometers (950 miles), and Iapetus with one of 1,435 kilometers (890 miles). These fill places eighteen, nineteen, twenty, and twenty-one respectively.

Then comes a surprise. In 1978, it was discovered that Pluto has a satellite, Charon. This satellite turns out to have a diameter of 1,200 kilometers (745 miles) so that it goes into twenty-second place. Its diameter is just about half that of Pluto, so that Pluto-Charon is a much better candidate for a double planet than Earth-Moon is. (Sorry, folks.)

After this, there come four more satellites. There are two of Uranus's satellites: Umbriel, with a diameter of 1,190 kilometers (740 miles), and Ariel at 1,160 kilometers (720 miles). Then there are two of Saturn's satellites: Dione at 1,120 kilometers (695 miles) and Tethys at 1,048 kilometers (740 miles). These four fill places twenty-three, twenty-four, twenty-five, and twenty-six, respectively.

That brings us back to the planets. There are tens of thousands of small planetary bodies circling the Sun, mainly between the orbits of Mars and Jupiter. These are the "asteroids" and the largest of them is Ceres, which, with a diameter of 940 kilometers (585 miles), fits into twenty-seventh place.

Beyond that, there are many satellites, asteroids, comets, and miscellaneous bits of matter that we can place in order. Let us stop with Ceres, then, and for convenience' sake, here is a tabulation of the results:

OBJECT	DIAMETER		
	Kilometers	Miles	Moon = 1
1. Sun	1,394,000	866,000	401
2. Jupiter	142,800	88,700	41
3. Saturn	120,600	75,000	34.7

4.	Neptune	50,200	31,200	14.4
5.	Uranus	49,000	30,440	14.0
6.	Earth	12,756	7,926	3.67
7.	Venus	12,140	7,544	3.50
8.	Mars	6,790	4,219	1.95
9.	Ganymede	5,262	3,270	1.51
10.	Titan	5,150	3,200	1.48
11.	Mercury	4,878	3,031	1.40
12.	Callisto	4,800	2,980	1.38
13.	Io	3,630	2,260	1.05
14.	Moon	3,475	2,160	1.00
15.	Europa	3,138	1,950	0.90
16.	Triton	2,735	1,700	0.78
17.	Pluto	2,500	1,550	0.72
18.	Titania	1,610	1,000	0.46
19.	Oberon	1,550	960	0.444
20.	Rhea	1,530	950	0.440
21.	Iapetus	1,435	890	0.412
22.	Charon	1,200	745	0.345
23.	Umbriel	1,190	740	0.343
24.	Ariel	1,160	720	0.333
25.	Dione	1,120	695	0.322
26.	Tethys	1,048	650	0.301
27.	Ceres	940	585	0.271

What's the good of this list? Well, it's pretty, and I don't know that it exists precisely in this form anywhere else. I like things that are pretty, and orderly, and different. After all, as I told you, I'm peculiar.

6
The Nearest Star

You would think that a person like myself is immune to frustration where the writing of books is concerned. The total number of my published books stands, at the moment, at 459. Therefore, it might seem that there isn't a book that I would like to write that I haven't written.

Yet it's not so.

In the 1970s, I wrote a series of astronomy books for the general public, each of them being full of tables and figures. The titles of four of them are (1) *Jupiter, the Largest Planet*, (3) *Mars, the Red Planet*. (4) *Saturn and Beyond*, and (5) *Venus: Near Neighbor of the Sun*. In these four, I managed to cover every planet in the Solar system and to say a few words about their satellites, the asteroids, and the comets.

The second book in the series is *Alpha Centauri: the Nearest Star*.

Well, Alpha Centauri is *not* the nearest star. Our Sun is. So I was planning to write a sixth book in the series that would deal with the Sun, both to round out my description of the Solar system and to correct the mistitling of my book on Alpha Centauri. I even have a contract, somewhere, for that book.

But it never got written. Other books intervened. They still intervene.

The next best thing, then, is to write essays on the Sun for this series, and that I now plan to do. Naturally, if this were a book, I would start with the apparent motion of the Sun in the heavens and talk about how this affects day and night, and the seasons, and the calendar, and astrology, and so on. However, I have written essays earlier in this series on all these subjects.

What I want to do now, then, is to write about the nature and the properties of the Sun, and I'll begin by asking, "How near is the nearest star?" Are you all ready? Good, let's go.

To all appearances, the Sun is simply a circle of light that moves across the sky in the course of the day. If anyone were to guess, purely from appearances, how large it was, the most popular guess would probably be, "Oh, about a foot across."

That can't be so, however. In order for an object a foot across to look the size of the Sun in our sky, it has to be only 114 feet away, and you know the Sun has to be farther away than *that*. So how far is it, and how big is it?

The first person we know of by name who tried to answer the question was a Greek astronomer, Aristarchus of Samos (310–230 B.C.).

Aristarchus knew that the Moon shines by the reflected light of the Sun and that at the moment when the Moon is exactly half illuminated, the Earth, the Moon, and the Sun should be at the apexes of a right triangle. If he could calculate the size of even one of the acute angles of the triangle, he could, by trigonometry, work out the relative lengths of the sides of the triangle.

87

His mathematics was completely correct, but he had no instrument with which to measure the angle precisely. His estimate was rather far off and he concluded that the Sun is about twenty times as far from the Earth as the Moon is. This was a radical underestimate, but given the circumstances, it was a remarkable achievement, so hats off to Aristarchus.

He went further. He studied eclipses of the Moon and realized that the curve of the darkness that invades the Moon at this time is the shadow of the curve of the spherical Earth. By comparing the curve of the Earth's shadow with the curve of the Moon's edge, he estimated that the Moon must be about one third the diameter of the Earth. Not bad!

And now he went still further. Since the Moon and the Sun are the same in apparent size, the real size must be proportional to the distance from Earth. If the Sun is twenty times as far as the Moon is, it must be twenty times as large as the Moon is. If the Moon is one third the diameter of the Earth, then the Sun must be about seven times the diameter of the Earth.

But that raises an interesting question. Why should a huge, bloated body like the Sun revolve about the midget Earth? Indeed, Aristarchus answered that question by suggesting that the Earth and the other planets move about the Sun.

It was really a remarkable bit of ratiocination that had only the tiny flaw of being eighteen centuries ahead of its time—because everyone laughed at him. Aristarchus's writings didn't survive and we only know of his suggestions because Archimedes (287–212 B.C.), the greatest of all the ancient mathematicians and scientists, mentioned them—and laughed.

The reason that Aristarchus couldn't put across his notions was twofold. First, it stood to reason that the Earth is *not* moving because any idiot can plainly see that it isn't. If you're on a horse, you can tell if it's standing still or if it is moving, even with your eyes closed, and shouldn't that be true if you're on Earth?

Besides, as Aristotle (384–322 B.C.) had earlier pointed out, if the Earth moved about the Sun, the motion ought to be reflected in a corresponding motion of the stars in reverse. (He was absolutely correct in this, but it didn't occur to him that the stars might be so far off that the stellar motion would be too small to measure.) Since the stars were clearly seen to be motionless, the Earth was motionless as well.

Suppose we forget about the motion of the Earth, however. Aristarchus merely took that as an inference from his work on the relative sizes of the Earth, Moon, and Sun. Even if the Earth is standing still, his calculations remain supported by impeccable mathematics (if highly peccable measurements). Why couldn't that be accepted?

Well, two centuries before, the philosopher Anaxagoras of Clazomenae (500–428 B.C.), while living in Athens, had suggested that the Sun is a blazing rock about a hundred miles across, and for this piece of blasphemy, among others, he was tried for impiety and atheism, and he thought it wise to leave Athens.

Of course, Anaxagoras was simply making an assertion, whereas Aristarchus had worked it out mathematically, and had come out with a size far huger than the earlier one. However, Cleanthes of Assoa, a Stoic philosopher who lived in Aristarchus's times (and who may have reached the age of 100 before he died), was fiercely offended and suggested that Aristarchus be tried for impiety, too. (As far as we know, though, he wasn't.)

Still, you can see that talk concerning the Sun's size and the Earth's motion wasn't encouraged even in the days when Greeks believed in free speech (more or less). It certainly wasn't encouraged when such freedoms disappeared in post-Greek times.

However, let's continue. About 240 B.C., the Greek philosopher Eratosthenes of Cyrene (276–196 B.C.) noticed the difference in the comparative lengths of shadows at different spots on the Earth. He attributed this, correctly, to the fact that Earth's surface is curved and he worked out a correct trigonometric method for calculating from that the size of the Earth. Since his measurements were good, he got an approximately correct answer. Rather than try to work out the figures Eratosthenes got, I will give you the current figures. The Earth has a diameter of 12,756 kilometers (7,926 miles) and a circumference of 40,075 kilometers (24,900 miles).

Furthermore, about 150 B.C., the greatest of all the Greek astronomers, Hipparchus of Nicaea (190–120 B.C.), attempted to measure the distance of the Moon and decided that its distance is equal to sixty times the radius of the Earth, or thirty times its diameter. This is just about correct and the current value of the average distance of the Moon is 384,401 kilometers (238,861 miles) or 30.13 times the diameter of the Earth.

If we stick to Aristarchus's suggestion that the Sun is twenty times as far from us as the Moon is, then the figures given us by Eratosthenes and Hipparchus tell us that the Sun is about 7,688,000 kilometers (4,777,000 miles) from us.

From the Moon's *apparent* size and its known distance, we can calculate that its absolute diameter must be 3,476 kilometers (2,160 miles). If the Sun has a diameter 20 times as large, it is 69,500 kilometers across (43,200 miles).

In other words, Greek astronomy ends by giving us a picture of a Sun at least 70,000 kilometers across, but for some 1,700 years people managed to ignore that. It was too unsettling.

Of course, one might argue that even if the Sun were that huge, it was still merely an insubstantial sphere of light and it was the Earth that, alone of all the objects in the Universe, was solid and heavy, so that it was the natural candidate for being at the center.

That brings us to the Polish astronomer Nicolas Copernicus (1473–1543), who returned to Aristarchus's idea of the Earth going around the Sun. He didn't come to that as a result of thinking about the huge size of the Sun, but only because it seemed to him that the mathematics of predicting planetary motions across the sky would be simplified if it was assumed that it is the Earth and the other planets moving about the Sun, rather than the Sun and the other planets moving about the Earth.

For years, Copernicus refrained from actually publishing his work lest he, like Anaxagoras and Aristarchus, get in trouble with the authorities. He finally did have it printed in 1543, but only when he was safely on his deathbed.

And at that, the publisher (who didn't want to get into trouble either) put in a preface stating that the Copernican theory was not intended to indicate that the Earth *really* moves about the Sun, but that it is only a mathematical device for calculating planetary motions.

It took an entire century for even the scientific and scholarly world to accept Copernicus. Only a few hundred copies of the first edition were printed. A second edition didn't appear till 1566 in Basel, and a third not till 1617 in Amsterdam. (In the original manuscript, by the way, Copernicus mentioned Aristarchus, then crossed it out. Either he didn't want to get into further trouble by reminding the world of that infamous crackpot, or he didn't want to share the credit, probably the former in my view.)

The English philosopher Francis Bacon (1561–1626), an early and strong advocate of experimental science, would not accept Copernicus because Bacon couldn't make himself believe that the huge, massive Earth is flying through space. Harvard University, which was founded in 1636, taught for years that the Sun goes around the Earth.

In 1807, when Napoleon's conquering career brought him to Poland, he visited the house where Copernicus was born and expressed his surprise that no statue had been raised in his honor. But it wasn't till 1835 that Copernicus's book was taken off the Index, which listed those books banned by the Roman Catholic Church. And in 1839, when a statue to Copernicus was *finally* unveiled in Warsaw, no Catholic priest would officiate at the ceremonies.

And even *today*, when a poll was taken of American *adults* as to the astronomical facts of life, 21 percent gave it as their opinion that the Sun goes around the Earth. Another 7 percent didn't know, or possibly didn't care.

How can we account for all the trouble Copernicus feared, for the length of time it took to accept him, and for the refusal of one quarter of American adults to accept him *today*?

My own feeling is that it all comes down to two verses in the Bible. Joshua was fighting the battle of Gibeon and it looked as though the enemy might escape under cover of night. Joshua 10:12–13 tells the story:

"Then spake Joshua to the Lord in the day when the Lord delivered up the Amorites before the children of Israel, and he said in the sight of Israel, Sun, stand thou still upon Gibeon. . . . And the sun stood still . . . until the people had avenged themselves upon their enemies. . . . the sun stood still in the midst of heaven, and hasted not to go down about a whole day."

Now how could Joshua (and God) possibly have ordered the Sun to stand still, unless it were ordinarily moving?

To those Fundamentalists of all periods who considered the Bible to be inspired by God with every word divinely true, it would seem on the basis of these verses (very famous ones) that the Bible says the Sun is moving,

92

and therefore going around the Earth, and that, frankly, I think explains everything.

You might argue, of course, that Copernicus's heliocentric theory is "just a theory," just a mathematical device.

But then came the Italian scientist Galileo Galilei (1564–1642) and his telescope. In January 1610, Galileo spotted four small objects (we now call them satellites) that move around Jupiter exactly as Copernicus said that planets, including the Earth, move about the Sun. For one thing, that showed that not *all* bodies move about the Earth. For another, it showed four small bodies moving about a much larger body. If the Sun were really much larger than Earth, as Aristarchus had shown, then it made sense for the Earth to move around the Sun.

However, there were arguments against Galileo's discovery. One scholar pointed out that since Aristotle hadn't mentioned these satellites anywhere in his writings, they didn't exist. Another pointed out that since they could be seen only through a telescope, they were the creation of the telescope, and didn't really exist. Others simply refused to look, and therefore didn't see them, which showed they didn't exist.

But suppose the satellites did exist. If Jupiter moved around the Earth, then it carried the satellites with them, and the satellites also went around the Earth. Then, too, since Jupiter and the satellites (and the Sun) were all made of immaterial heaven stuff, and only Earth was solid and heavy, it didn't matter how the heavenly bodies went around each other, they all, in the end, went around the Earth.

But then came the matter of Venus. It reflects the light from the Sun, and if the notion of Venus and the Sun both going around the Earth were correct, Venus should always display a crescent. If, on the other hand, Venus and the Earth were both going around the Sun, Venus should show a full complement of phases, exactly as the Moon does.

Galileo studied Venus through his telescope and, by December 11, 1610, was satisfied that the phases are a full complement, like the Moon. That put him into a quandary. Should he announce that the phases of Venus show that the Earth moves around the Sun, and not vice versa, and get himself into deep trouble? Or should he keep quiet and risk losing credit for the discovery? (Other people were by now using telescopes, too.)

What Galileo did was to publish a Latin phrase, therefore, that went: "Haec immatura a me iam frustra legunter o.y." This means "These unripe things are read by me." The "o.y." was added only to make it come out right. Then, if anyone tried to claim later credit for the discovery, Galileo needed only to rearrange the letters anagram-wise, and get "Cynthiae figuras aemulatur Mater Amorum." This is Latin for "The Mother of Love imitates Cynthia's shape," where the Mother of Love is clearly Venus and Cynthia is one of the poetic names for the Moon.

But how could one defeat the argument that heavenly bodies are so different from Earth, you can't reason from one to the other. The first thing that Galileo had looked at with his telescope, in 1609, was the Moon, and he had made out mountains, and craters, and seas on its surface. The Moon might be a heavenly body, but it is clearly an Earthlike world.

What about the fact that heavenly bodies glow with light and the Earth is dark? Well, the planets *don't* glow with light. The fact that the Moon and Venus show phases demonstrates that they are dark bodies shining by the reflected light of the Sun. What's more, Galileo argued, the dark part of the Moon, when only a crescent is shining, can be made out by a very soft light, and that must be the result of Earth shining by reflected sunlight in the Moon's sky, as the Moon shines in ours. We are seeing "Earthshine" on the Moon, so that Earth, too, glows with light as the Moon and Venus do.

Leonardo da Vinci (1452–1519) had noted Earthshine

and interpreted it correctly over a century before, but he had saved himself a lot of trouble by not publishing. Galileo was less cautious. In 1632, he published all his reasoning in "Dialogue on the Two Chief World Systems," and did it in Italian, so that all his countrymen could read it, and not just a bunch of effete scholars who understood Latin.

You all know what happened. In 1633, Galileo was called before the Inquisition, threatened with torture, and forced to state that the Earth does not move, and to promise nevermore to say that it does.

There is a legend that as Galileo left the court, he muttered under his breath, "Eppur si muove!" ("But it moves just the same!"), referring to Earth. He may not have. He was nearly seventy years old and playing the wise guy with the Inquisition, which was notorious for its lack of a sense of humor, would have been incredibly foolish.

But as a matter of fact, the Earth *did* move anyway.

This is not to say that Copernicus was completely right. The heliocentric theory is *not* "just a theory." It has innumerable observations and extraordinarily close reasoning to back it up, as all good theories must have. But that doesn't mean that Copernicus had it right in every detail. It could be improved. (Scientific theories can always be improved and *are* improved. That is one of the glories of science. It is the authoritarian view of the Universe that is frozen in stone and cannot be changed, so that once it is wrong, it is wrong forever.)

Thus, Copernicus, in switching from an Earth-centered planetary system to a Sun-centered one, kept the old Greek idea that the bodies move about the center in perfect circles or combinations of circles.

The German astronomer Johann Kepler (1571–1630), studying the best observations of Mars's positions ever taken up to that time—by the Danish astronomer Tycho Brahe (1546–1601)—realized that a circle wouldn't do. Mars, and presumably all the planets, move in elliptical orbits about the Sun, with the Sun not at the center of the ellipse, but at a focus a little to one side of the center. He also worked out the changing speeds at which planets had to move in such orbits, and how those speeds change with distance from the Sun.

It all boiled down to Kepler's "Three Laws of Planetary Motion," published in 1609 and 1619. They have not had to be measurably modified in all the nearly four centuries since, so Kepler finally got the Solar system right.

It was possible now to make a model of the Solar system with the Sun in the center and with all the planetary orbits about it to exact scale. (At least one can do it mathematically, and even take into account how each planet moves in its orbit, and therefore, where at any given moment, all the planets would be relative to each other and to the Sun.

The Moon is not part of this model because it goes around the Earth and not around the Sun, so that it doesn't fit into Kepler's laws the same way. That goes for Jupiter's satellites, too.

Therefore, if the distance of any planet from the Earth is determined, then the distance of all the other planets *and the Sun* can be easily calculated from the Keplerian model.

This offers two advantages. At least three of the planets—Mars, Venus, and Mercury—are closer to the Earth than the Sun is, at least some of the time, and their distance should be more easily measured. And it is easier to make exact measurements of the pointlike planets than of the large globe of the Sun, which is unbearable (and blinding) to look at.

Of the three planets, Mars is the easiest to observe because it is often present in the sky during the whole night, whereas Venus and Mercury are only present in the evening and in the dawn. So how do we determine the distance of Mars?

One way to do it is to measure the position of Mars against some nearby star, first from one place on Earth's surface, and then from another place on Earth's surface either at the same time or at a known difference in time. If this is done, Mars's apparent position shifts relative to the star, provided the star is much farther away than Mars is. This is called "parallax."

You can see how parallax works if you hold your finger up and note its position against objects on the wall of the room. View with your left eye only, then your right eye only. The finger shifts position. So do the objects on the wall of the room, but since they are farther away, they shift less, and you are much more aware of the finger's shift.

A star is so distant that it doesn't seem to shift at all and it can be considered the immovable background. However, Mars is also so far away that its parallax is ordinarily too small to be observed.

What do we do, then? For one thing, the greater the change in position from which an object it viewed, the greater the parallax. That means you should view Mars from two places that are thousands of miles apart. But you've got to view it from those two places at known times so that you have to have good clocks that are synchronized, either against each other, or against the time when a certain star crosses the meridian. Even then, the parallax is very small and you must have a telescope with which to make the measurements.

Everything came together in 1671. By that time, telescopes were much improved over Galileo's first instruments. An adequate method of telling time had come about with the invention, in 1656, of the pendulum clock by the Dutch scientist Christiaan Huygens (1629–1695). And to top it off, two astronomers made measurements thousands of miles apart.

The Italian-French astronomer Giovanni Domenico Cassini (1625–1712) was observing Mars in Paris. In Cayenne, French Guiana, on the other side of the Atlantic Ocean, a French astronomer, Jean Richer (1630–1696), was also observing Mars, at Cassini's direction.

The distance between Paris and Cayenne, in a straight line passing through the bulge of spherical Earth, could be calculated. From the difference in position of Mars relative to the nearby stars, the distance of Mars could be calculated by Cassini. From the distance of Mars at that time, at that place in its orbit, the distance of the Sun could be calculated.

Cassini did all this, and for the first time in the history of the world, the distance of the Sun was calculated with reasonable accuracy. Cassini actually came out some 7 percent low, but for a first attempt, that was terrific. Let me give you the distance by present standards. Making use of parallax methods under improved circumstances, and other devices that only became practical in the last few decades, we now know quite certainly that the Sun is about 149,600,000 kilometers (92,960,000 miles) from the Earth. This is 19.5 times as great as Aristarchus's figure.

Now we know how near the nearest star is, so we have the answer to our first question.

If the Sun is that much farther away than Aristarchus had thought, it must also be that much larger than Aristarchus had thought, for, whatever its distance, its *apparent* size in the sky doesn't change.

We now know that the diameter of the Sun is 1,394,000 kilometers (866,000 miles). This means it has a diameter that is 109 times that of Earth.

It also means that it has a volume nearly 1,300,000 times that of Earth, so that if the Sun were hollow you could drop into it 1,300,000 bodies the size of the Earth, provided you ground them all up into dust.

If Aristarchus had been able to make his measurements accurately, then by the time Hipparchus had done his work, the true diameter of the Sun would have been known and it would really have been much easier to suppose the Earth is moving around the Sun.

Except that people might still have argued that no matter how big, how huge, how impossibly colossal the Sun is, it is still simply an object of immaterial light and weighs nothing and therefore it must carry its vast bulk about the Earth.

This is something we must take care of. How can we show that the Sun and the other heavenly bodies are *not* immaterial; that they are as heavy and as massive as Earth is, if not even more so?

In short, we must go on to the next question: How much does the Sun weigh? Or more appropriately: What is its mass?

We'll carry on in the next essay on this question.

7
Massing the Sun

When I was young, I read a great deal of poetry. Partly, that was because poetry was pushed at me in school. (I don't know if it is anymore, but I certainly hope it is.) And partly it was because I didn't know any better. My immigrant parents, as I have frequently explained, did not know enough about English literature to guide my reading, so I read *everything*. I even read stuff like poetry, which children were supposed to hate, because no one told me I was supposed to hate it.

In any case, I remember much of the poetry I read in those days because I have always had trouble forgetting anything (except things that are vital, like the instructions my dear wife, Janet, gives me, in her hopeless attempt to make me live forever). And some of the poetry has persisted in coloring my view of the world even today.

For instance, there is a poem by Francis William Bourdillon (I won't lie to you—I had to look up his name), who wrote a short poem, of which the first verse goes as follows:

> *The night has a thousand eyes*
> *And the day but one;*
> *Yet the light of the bright world dies*
> *With the dying sun.*

Nothing I have read, either in literature or in science, has given me so unfailing an appreciation of the importance of the Sun as those four lines.

As a result, I was not the least surprised to find, a little later in my youth, that the first monotheist we know of in ordinary secular history, who was the Egyptian pharaoh Akhenaton (reigning 1372–1362 B.C.), chose the Sun as the one supreme God. (Good choice, Akhenaton, I thought. Very logical.)

So I will continue with my discussion of the Sun. In the previous essay, I posed the question: How far is the Sun? The answer turned out to be 149.6 million kilometers (92.96 million miles), and at that distance, the Sun turns out to be 1,394,000 kilometers (866,000 miles) in diameter.

Let us now ask a second question, which I will put in a deliberately naive form: How heavy is the Sun?

The conventional view of Western religion was, to begin with, that the Sun was merely a container of light.

Light itself was the first thing mentioned in the Bible as having been created, for on the first day, "God said, Let there be light: and there was light." (Genesis 1:3.)

To begin with, we might assume, light simply permeated the Universe, but (still on the first day), God separated light from dark to make both day and night.

It was not till the fourth day that the light was collected into the various heavenly objects, of which incomparably the brightest is the Sun. The Moon is a very distant second, and the stars are just little sparkles. "And God made two great lights; the greater light to rule the day, and the lesser light to rule the night: he made the stars also." (Genesis 1:16.)

There existed light on Earth, too, independent of the heavenly bodies. There were brief, occasional lightning strokes that could start forest fires. And eventually, there were fires that could be brought into existence by human beings, yielding light and warmth.

The study of Earthly sources of light made an important point very clear. It seemed obvious that light has no weight. It is immaterial.

From this, an important conclusion could be drawn. If the Sun were nothing more than a ball of light, it, too, must be immaterial and have no weight. If so, it wouldn't matter, if it was very far away and, therefore, very huge. No matter how large it was in sheer size, it would still be a weightless bit of less-than-fluff, and it might be argued that it would then have to circle the heavy Earth.

A second basic observation of light on Earth is that it cannot exist for long unless the fire that emits it is constantly fed on fuel. Any source of light dies out as soon as the wood or oil is consumed.

The Sun, on the other hand, does not go out. It has emitted light, unchanged, all through human history and shows no sign even today of diminishing or fading, let alone going out. Nor is there any sign that, in the process, the Sun is consuming fuel.

The conclusion can be drawn from this that Earthly light has certain basic differences from heavenly light. Earthly light is a temporary phenomenon based on fuel; heavenly light is eternal and requires no fuel. This is an important example of the seeming fact that the laws of nature on Earth are different from those in heaven.

The Greek philosopher Aristotle (384–322 B.C.) reasoned this out in considerable detail. In general, Earthly objects are imperfect and time-bound; decay and corruption are inevitable. Heavenly objects, on the other hand, are eternal and incorruptible; in short, perfect. Then, too, Earthly objects, if left to themselves, do not move or, if they move, do so by falling downward or rising upward. Earth and water tend to sink; air and fire to rise (these being the four basic substances or "elements" that make up the Earth in the Greek view). Heavenly objects, on the other hand, always move, and in doing so, neither fall nor rise but progress in grand, unchanging circles about the Earth.

Since that is so, heavenly bodies cannot be made out of earth, water, air, or fire, but out of some different substance altogether, which Aristotle called "aither" and we call "aether" or "ether." It is from the Greek word for "blazing," since the heavenly objects glow with an eternal light, whereas Earthly objects are dull and dark, except for occasional bits of human-made fire that are very imperfect when compared to the divine fire of the ether.

Rocks and heavy objects, generally, fall unless supported. We all know that. Hold a rock out and let go and, instantly, of its own accord, it falls. But why?

To answer that, Aristotle suggested that every object has a natural place in the Universe, and if it is outside that natural place, it makes every effort to return to it, provided it is not constrained. As long as you hold the rock in midair, it is constrained to remain there, but you feel its weight as it struggles, so to speak, to plunge toward the center of the Universe, which is the natural place for solid substances. And if you let go, it instantly moves toward that center; in other words, it falls.

It seemed to Aristotle that the heaviness of an object is the measure of the intensity of its longing to be in its natural place. Therefore, a heavy object would naturally fall more quickly than a light object. A rock would fall more rapidly than a leaf, and a leaf would fall more rapidly than a small, downy feather.

You can, if you wish, easily satisfy yourself that Aristotle was wrong in this view by a simple experiment:

Take two identical sheets of paper and drop them simultaneously. They will both fall rather slowly and at equal speeds as they do so. Now take one of those sheets of paper and crumple it into as small a ball as you can manage. Its weight has not changed in the process, so now you are ready to drop two objects, one thin and flat, and one crumpled into a small compact structure, but both of which are the same weight.

Drop them and, behold, the crumpled paper falls considerably faster than the flat paper of equal weight.

Why? Because materials falling in our atmosphere have to push air molecules out of the way as they fall and that consumes some of the energy of falling and makes them drop more slowly. If an object is quite heavy, this slowing through air resistance is negligible, but it becomes greater if the object is light. It becomes greater still if the object is both light and presents a large surface to the air.

To us, in hindsight all this seems obvious, but it wasn't till the time of the Italian scientist Galileo (1564–1642), nineteen centuries after Aristotle, that the old Greek's notions of falling bodies were actually put to the test.

In the 1590s, Galileo did two things that were important. First, he used objects that were so heavy that air resistance was negligible. Second, he let the objects roll down an inclined plane, which diluted and slowed the natural tendency to fall so that he could more easily observe and measure the speed with which they moved.

His experiments showed one all-important fact—that objects, regardless of their weight, all fall at the same rate. Legend has it that Galileo demonstrated this by simultaneously dropping two heavy balls, one ten times heavier than the other, from the top of the Leaning Tower of Pisa, and having them hit the ground—

thunk—simultaneously. It is almost certain that he didn't do this, but his experiments with balls rolling down an inclined plane did the job just as well, if less spectacularly. They served to kill Aristotelian physics.

(In a vacuum, where air resistance is absent, it would follow that *all* objects, however light and however large the surface, fall at the same rate. In a vacuum, a feather would fall as rapidly as a cannonball. This has been tested, and it is so.)

About three quarters of a century later, the English scientist Isaac Newton (1642–1727) took the findings of Galileo and others in connection with moving objects, and worked out three assumptions called the "Three Laws of Motion," which satisfactorily explain all the varieties of motion encountered on Earth.

The Second Law of Motion states that if a force is exerted on a body, that body will undergo an acceleration. It will speed up, or slow down, or change the direction of travel, depending on the direction in which the force is exerted. What's more, the same degree of force will produce a smaller acceleration in a heavier object than in a lighter one. (To see that this is so, first kick a football and then kick a cannonball in just the same way, and see what happens.)

Newton defined an innate property of matter, which he called "mass." The greater the amount of mass a particular object has, the less it is accelerated by a given force. On the surface of the Earth, the mass of an object is proportional to its weight, but the two are not identical. Weight changes with position in the Universe, but mass does not—something we needn't go into at this point.

Newton's law of motion have held for all ordinary conditions ever since, and they have proved a completely satisfactory way of dealing with motion. (Under extreme conditions, Einstein's generalization of these laws is more useful, but we won't go into that at this point, either.)

The replacement of Aristotle's ideas of motion by those of Galileo and Newton did not, in themselves, necessarily alter the proposition that the laws of nature are different on Earth and in the heaven. In whatever manner you explain the way in which bodies fall on Earth, they do fall; and the heavenly bodies do not fall, but move in circles.

They seem to move in circles around the Earth, but even if heliocentric views, as described in the previous essay, are accepted, and if some heavenly bodies move around the Sun, they still move in circles and don't fall.

Now what? Well, let's take a different tack.

The ancient Greeks thought that the planets move in circles, not because the planetary motions indicated that, but because a circle was viewed as the simplest and most perfect curved figure. In their eyes, nothing less than perfection would do for the heavenly bodies.

Since the planetary motions do not move as they would if their orbits were perfect circles, the Greeks supposed that they travel in combinations of circles and build up structures of circles upon circles that grow more and more complicated. They insisted on forcing the actual motion of the planets into their concept of what was neat and pretty. (They called it "saving the appearances.")

When Copernicus, as described in the previous essay, decided that the planets, other than the Moon, are actually moving about the Sun, he still felt that they travel in combinations of circles. He couldn't get rid of that particular Greek idea.

The one who broke the spell was a German astronomer, Johann Kepler (1571–1630). He had a set of observations of Mars that had been recorded by a Danish astronomer, Tycho Brahe (1546–1601), which were the best ever made up to that time. Kepler tried to match those positions to a circular orbit, but could not make that work.

Desperately, he tried other types of curves and found that the ellipse (a slightly flattened circle) fit the orbit quite exactly. He therefore worked out the "Three Laws of Planetary Motion," the first two in 1609, and the third in 1619.

The first law states that a planetary orbit is an ellipse. An ellipse has a center, as a circle has, but an ellipse also has two foci, one on either side of the center, and the Sun is located at one of the foci of the planetary orbital ellipse, not at the center.

The second law describes how the speed of motion of a given planet changes with changing distance from the Sun. (With the Sun at one focus of the planetary ellipse, the distance between the Sun and the planet changes as the planet progresses along its orbit.)

The third law describes how the length of time different planets take to move about the Sun is related to their distances from the Sun.

Kepler was the first to describe the Solar system essentially as it is, and his description holds today as tightly as it did in his own time (allowing for some refinements due to Newton and, later, to Einstein). It is unlikely that the picture of the Solar system will ever have to be changed substantially from that given us by Kepler.

The trouble is, though, that the Three Laws of Planetary Motion are quite different from the Three Laws of (Earthly) Motion, so that it might *still* seem, even as late as the mid-1600s, that the laws of nature were different on Earth and in the sky.

In 1666, however, Isaac Newton (1642–1727) had left London, which was being decimated by the plague, and retired to his mother's farm where, one evening, he happened to watch an apple fall from a tree (and no! it did *not* hit him on the head) at a time when the Moon was shining in the sky.

He wondered why the apple fell and the Moon does not, then it occurred to him that the Moon *is* falling, but that it is also moving sideways and the combination of the two motions keeps it in orbit about the Earth. From the nature of the orbit, one could calculate how far the Moon falls toward Earth in one second and it is falling much more slowly than the apple was. Of course, the Moon is much farther from the Earth than the apple was and perhaps the attractive power of the Earth decreases with distance.

Since the intensity of light was known to decrease as the square of the distance, the intensity of Earth's attractive pull might decrease in the same way. Newton did the calculations and ended with a rate of fall for the Moon of only seven eighths of what it really was. That seemed to kill his theory, and he abandoned it in disappointment.

Why did his calculations fail? For one thing, he was using a figure for the radius of the Earth that was substantially smaller than the truth. That affected his calculations. It was also true that different parts of the Earth might attract the Moon at slightly different distances from slightly different directions, and Newton wasn't sure he knew how to allow for that.

Then one day, in 1684, English scientists were discussing the possibility that planetary motions were controlled by the Sun's attraction and Newton's friend Edmund Halley (1656–1742) asked him what path a planet would take about the Sun if the attractive power of the Sun decreased as the square of the distance.

"An ellipse," said Newton.

"How do you know?" said Halley.

"Why, I once calculated it," said Newton.

Halley grew fearfully excited, and when he found that Newton had come out with the wrong figure, he insisted that he try it again. By now, Newton had invented the calculus, which gave him just the mathematical tool he needed for his calculations. What's more, the French astronomer Jean Picard (1620–1682) had, in 1671, published a new estimate of the radius of the Earth that was more accurate than the one Newton had used in 1666.

When Newton went back to his calculations, he could see that they were now coming out *right* and he had to stop and recover before the excitement of it gave him a stroke.

Halley then drove Newton on mercilessly to write a book describing his laws of motion and all that could be deduced from them. Halley read the proofs, and undertook the expense of publication (he was rich). The book, the short title of which is *Principia Mathematica*, was published in 1687 and, by general agreement, is the greatest work of science ever presented to the world.

Newton worked out the Law of Gravitation from his Three Laws of Motion. He went on to work out the Three Laws of Planetary Motion from his Law of Gravitation.

Mind you, Newton did not just propose a simple law of gravitation. Even primitive man knew that all heavy objects fall to Earth. Everyone knew that Earth is the source of an attractive pull. It took no Newton to tell the world that.

What Newton proposed was the Law of *Universal* Gravitation. Every particle in the Universe that possesses mass attracts every other particle in the Universe that possesses mass, and does so with a force proportional to the product of the two masses and inversely proportional to the square of the distance between them.

That made it possible to determine the relative masses of different objects in the Universe. For instance, the Moon doesn't simply revolve about the Earth. By Newton's law, both Earth and Moon revolve about a common center of gravity. This center of gravity is on a line connecting the centers of the Earth and the Moon and its distance from the Earth's center and the Moon's center is inversely proportional to the masses of each. The center of gravity can be located because, as the Earth goes around it, the stars seem to wobble slightly in the course of a month.

The center of gravity of the Earth-Moon system is

(at an average) 4,728 kilometers (2,938 miles) from the center of the Earth. This is actually 1,650 kilometers (1,025 miles) under the Earth's surface, so it's not a bad approximation to say that the Moon is orbiting the Earth.

The center of gravity is 81.3 times as close to the center of the Earth as it is to the center of the Moon. This means that the Earth's mass is 81.3 times that of the Moon's mass. We can't use the position of the center of gravity to tell us the absolute mass of either body, but we can get the relative masses of the two, and that's plenty.

And what about the mass of the Sun?

We know how rapidly the Moon goes around the Earth. If the Moon were farther away from Earth, it would have to travel over a longer orbit and it would also move more slowly since Earth's gravitational whip would be feebler. We could take both effects into account and calculate how fast the Moon would be moving and how long it would take it to circle the center of gravity of the Earth-Moon system at any distance—at a distance, for instance, equal to that separating the Sun from the Earth.

If the Moon were 149.6 million kilometers from Earth, and if there were no other heavenly bodies in the vicinity to interfere, it would be moving very slowly indeed, much more slowly than the Earth moves about the Sun at that very same distance.

Why does the Earth move much more quickly under the Sun's influence than the Moon would under Earth's at the same distance? Obviously, because the Sun's gravitational pull is much stronger than the Earth's.

And why is the Sun's pull much stronger than the Earth's? Because the Sun has more mass. From Newton's Law of Universal Gravitation, and from the known orbital speeds of the Moon and the Earth, together with the known distance of the Moon from the Earth and of the Earth from the Sun, we can calculate the Sun's mass relative to the Earth's.

It turns out that the Sun is no immaterial expanse of light. The Sun is a material body with a mass equal to 333,000 times that of the Earth, and at any given distance, the Sun's gravitational pull is 333,000 times the intensity of Earth's gravitational pull.

In other words, once the consequences of Newton's Law of Universal Gravitation were understood, there remained no longer any conceivable reason (other than blind and stubborn faith in nonsense) to suppose that the Sun is circling the Earth.

In 1798, the English scientist Henry Cavendish (1731–1810) measured the force of gravitation between two metal balls in his laboratory and from that calculated the actual mass of the Earth. From that we can get the absolute masses of the Moon and the Earth, but these are such huge figures we can't visualize them. If we stick to relative masses and suppose that the Moon's mass is 1, we have:

Mass of Moon = 1
Mass of Earth = 81
Mass of Sun = 27,000,000

Is there anything else we can tell about the Sun right now? Yes, we know that the Sun has a diameter equal to 109 times that of the Earth (see the previous essay) and therefore has a volume equal to $109 \times 109 \times 109$, or 1,295,000 times that of the Earth.

If, then, the Sun were made of precisely the same sort of material that the Earth is made of, it should have a mass equal to 1,295,000 times that of the Earth, but it doesn't.

Since the Sun's mass is only 333,000 times that of the Earth, it is made of (on the whole) lighter material than the Earth is made of. The density of the Sun (the number of kilograms per cubic meter, or the number of pounds per cubic foot) is 333,000/1,295,000, or only ¼ that of Earth.

Now we can come to a grand conclusion. We have shown that the Moon and Sun have mass as Earth has. By similar methods we can show that every material heavenly body has mass, even distant stars and galaxies. (Light and a few other substances are immaterial and don't have mass in the ordinary sense of the word, but we can ignore that.)

Again, since all the objects with mass in the Universe seem to obey the Law of Universal Gravitation in accordance with Newton's equations (barring extreme cases where we must use Einstein's generalization), it would appear that all bodies everywhere in the Universe, whether on Earth or in the most distant galaxy, obey the same laws of nature—all the laws of nature, we might guess, and not just the Law of Universal Gravitation.

This remains an assumption, of course, because we can't test the Universe directly at great distance, but we have not yet discovered anything in the three centuries since Newton's great book was published that would lead us to doubt, in any serious way, the universality of the laws of nature as determined on Earth.

And yet, even so, there can remain questions that are not answered by Newton's equations. For instance:

The Sun has mass, but Sun stuff is not exactly Earth stuff since Sun stuff has only a quarter of the density of Earth stuff.

Is this because the Sun is so much hotter than Earth? After all, density does tend to go down as the temperature rises.

Or is it because the Sun is composed of the same material that Earth is, but in different proportions? Some materials on Earth are considerably less dense than other materials and maybe the Sun is made up mostly of the less dense materials we know on Earth.

Or can it be that the Sun is made up of substances

utterly different from those making up the Earth? Even with all their obedience to the laws of nature, it may be that heavenly objects are fundamentally different in chemical composition than Earth is. Perhaps *every* heavenly object has its own composition, with no two alike.

And if that is so—or not so—how can we ever tell? We can't go to the Sun, for instance, and sample its material and analyze it.

In fact, the French philosopher Auguste Comte (1798–1857) stated, in 1835, that the chemical constitution of the stars was an example of the kind of information science would be eternally incapable of attaining.

Sometimes, though, these flat statements of "Impossible!" are dangerous (even though I, myself, make them often enough). Comte died just two years too soon to witness scientists learning how to obtain the kind of information he thought science would be eternally incapable of attaining.

In the next essay, then, we'll consider how that was done, and we'll find out what the Sun is made of.

8
What Are Little Stars Made of?

My dear wife, Janet, and I have a small ritual that consists of tracking each other. We are almost always together, but occasionally one or the other of us is forced to venture out into the wide world alone. Since we grow uneasy if we are at opposite ends of the apartment, you can well imagine that this separation is traumatic.

First, there are the preliminary anxious injunctions about "being careful" and references to madcap traffic, falling cornices, and dubious bystanders. Then, when either of us gets to where he or she is going, a phone call reporting safe arrival is considered de rigueur. And finally, an estimate of the return time having been made, we know when to put our worry machines into high gear. So far, I must say, nothing has happened to either of us, but each new time is a new risk.

Janet is particularly good at this and manages to begin being concerned about half an hour *before* the time I have set for return. I sometimes miss because I may get involved in my writing, and when that happens, I often, quite literally, don't know what time it is.

Janet, as it happens, goes out on many Monday evenings to meetings at her psychoanalytic institute, and invariably comes home between nine and nine-fifteen.

Then came one Monday when I was busily involved at the word processor and I happened to look at my watch and notice that it was 10 P.M. Janet had not yet returned, but I had forgotten about that. What clicked in my mind was that it was time for "Newhart," one of my favorite TV shows, so I turned it on.

At 10:05 P.M., Janet returned, having been kept late by some sort of prolonged discussion at the institute, and in a semifrenzy, thinking that I was half dead with worry, she was about to apologize feverishly when she noticed that my eyes were on the TV screen and that I was waving to her absently.

She said, rather sharply, "Weren't you *worried*?"

I'm an old hand at the marriage game, of course, so I knew better than to admit I had lost track of the time. I said, indignantly, "Of course, I was worried. Fearfully worried. Desperately worried."

"And what were you going to do about it?" she wanted to know.

"I was going to call the institute, ask where you were, and if you were still there, I was going to come down and get you."

She said, "*When* were you going to do that?"

And I said, pointing to the TV, "Just as soon as 'Newhart' was finished."

Fortunately, Janet has a sense of humor, so she burst into laughter and said she was glad to know her place in the scheme of things.

Well, after the two previous essays, I am still trying to place the Sun in the scheme of things, so let's continue.

* * *

As I explained in the previous two essays, it was Aristotle's opinion that the Sun, and all other heavenly objects, were composed of materials totally different from those making up the Earth, and for nearly two thousand years, scholars followed him in this thought.

Only in 1609 did this long-held opinion began to be shaken. In that year, Galileo (1564–1642) pointed a telescope at the sky for the first time. Looking at the Moon, he saw a world with craters and mountain ranges and what looked like seas. In short, it closely resembled Earth in physical appearance.

What's more, as other astronomers began to use telescopes, it was found that the planets showed disks, and were therefore worlds rather than points of light. They rotated on their axes as Earth did. Some showed clear evidence of atmosphere and clouds. Mars had polar ice caps, and so on.

The more planets were studied, the more they seemed Earthlike in one fashion or another, and if they had the *appearance* of Earth, might they not have the general chemical composition of Earth? Still, argument from similarity of appearance is interesting, but it's not proof.

But other types of similarity arose. Isaac Newton (1642–1727) published his theory of universal gravitation in 1687, and showed, quite conclusively, that the planets were subject to the same gravitational forces that Earthly objects were. He even considered those forces to be universal, governing every object in the Universe. If earthly and heavenly bodies were, after all, subject to the same laws of nature, might they not be composed of the same general substances?

Well, for one thing, Newton's view had its limitations. He might say that the law of gravitation was universal, but he could only apply it to the objects making up the Solar system. To be sure, in Newton's time, the Solar system was thought to make up the whole Universe except for an unimportant powdering of background stars on the firmament.

By 1793, however, the German-born British astronomer William Herschel (1738–1822) had shown that there were double stars that circled each other. These circlings were shown to fit Newton's theory of universal gravitation exactly, and since then, there has been no doubt that gravitation (and, indeed, the other laws of nature) are universal.

Nevertheless, even if everything in the Universe obeys the same laws of nature, that does not necessarily prove that all objects in the Universe are made up of the same basic materials. An ivory billiard ball and a plastic billiard ball may follow the same laws of motion and may have the same elasticity, hardness, and so on, but that doesn't prove that ivory and plastic are identical materials.

It would seem that the only way we could really determine the chemical nature of a heavenly body such as the Sun would be to actually obtain a piece of it and subject it to chemical analysis. It is not surprising, then, that in 1835 Auguste Comte (as I said at the end of the previous essay) gave it as his opinion that the chemical constitution of the stars was an example of the kind of information science would be eternally incapable of attaining.

But something *does* reach us from the stars, and from the Sun in particular. That fact has been known ever since humanity's ancestors developed a brain large enough to enable them to wonder about the surrounding Universe. Light reaches us from all the heavenly objects we can see, and most of all from the Sun.

The question, then, is whether light can tell us anything about the chemical constitution of those objects, in particular, of the Sun.

If light were uniform, structureless, and unchanging, no matter what the source, then it would yield us no information and would be useless as a guide to the nature of the source emitting it.

In 1665, Newton showed that light is *not* uniform and

117

structureless. It is a mixture of different colors that can be spread out, when passed through a glass prism, into a spectrum of colors: red, orange, yellow, green, blue, and violet, each one shading gradually into the next.

Light is obviously bent (that is, "refracted") through glass to different degrees, but the difference in properties that makes such differing refraction possible remained speculative until 1801, when a British physicist, Thomas Young (1773–1829), demonstrated that light is a wave phenomenon, and he even showed that the waves are tiny, less than a millionth of a meter in length. Presumably, then, ordinary light—say, from the Sun—is a mixture of waves of light of different length. The glass prism sorts them out, each bit of light being refracted to an amount that depends on its wavelength; the shorter the wavelength, the greater the refraction. The light spectrum, then, consists of an orderly progression of wavelengths from the longest visible waves of red light to the shortest visible waves of violet light, the different groups of wavelengths reacting differently with the pigments in the retina to give rise to the notion of color in the brain.

There was some question as to the nature of light waves. There are two kinds of waves known. One kind is like those in sound, where the waves are compressed and relaxed alternately in the direction of travel ("longitudinal waves"). There are also waves of the kind that appear on the surface of disturbed water, where the waves move up and down at right angles to the direction of travel ("transverse waves").

In 1814, the French physicist Augustin Jean Fresnel (1788–1827) showed that light waves were transverse waves.

Now how does that help us? The Sun sends out a vast mixture of tiny transverse waves, of every conceivable wavelength, and we can separate them and line them up in order, but how does one go from there to chemical composition? I doubt that anyone in Fresnel's times thought that there was any question of going from one to the other. There seemed no connection whatever.

In that same year of 1814, however, a German optician, Joseph von Fraunhofer (1787–1826), was working with lenses, prisms, and so on. He was the best optician of his time and he had to know the precise refracting abilities of the glass he used. He tested prisms by passing sunlight through them and forming a spectrum.

However, the spectrum was not continuous. It was interrupted by narrow regions of darkness. Here and there in sunlight there were missing wavelengths of light and therefore the dark regions.

There is a small puzzle as to why these dark interruptions were never noticed before. Some think that Newton, once he formed the first spectrum by use of a prism, ought to have noticed them. In 1802, a British chemist, William Hyde Wollaston (1766–1828), saw a few but, attaching little importance to the matter, didn't follow it up.

To me it seems simply a matter of Fraunhofer making use of very high-quality glass and perfectly ground prisms. Where Wollaston observed some seven regions of darkness, Fraunhofer detected nearly six hundred. What's more, he mapped them, noting that they always fell in the same spots along the spectrum, whether he used direct sunlight, or sunlight reflected by the Moon, or the planets. (Modern physicists, by the way, can detect some ten thousand lines in the solar spectrum.)

For over forty years, nothing much was done with those spectral lines, and yet chemists were aware of an interrelationship between earthly light and chemical composition.

As early as the 1750s, a Swedish mineralogist, Axel Fredrik Cronstedt (1722–1765), had introduced the use of a blow-pipe, in which a hot flame was played upon minerals. Information about their chemical constitution

119

could be obtained from the colors that appeared in the flame, in the vapors produced, in the ash, and so on.

As time went on, it was discovered that the vapors of heated substances produced a characteristic light. Hot sodium vapors tended to produce a yellow light; potassium vapors a violet light; barium vapors a green light; strontium vapors a red light; and so on. It is indeed such effects that are used to produce spectacular sprays of color in fireworks.

One of those interested in the study of these colors was a German chemist, Robert Wilhelm Bunsen (1811–1899). He studied heated minerals carefully, but of course, the flame he used for heating usually contributed colors of its own, which tended to obscure the results.

Bunsen, therefore, made use of a kind of burner, which he improved in 1857, that fed air into natural gas, producing such complete burning that a flame was produced that was hot indeed, but produced very little color. This was a "Bunsen burner," and when I was studying chemistry half a century ago, such burners were still absolutely essential pieces of laboratory equipment.

Using the Bunsen burner meant that the colors developed by heating minerals were clearer, less confused, and more effective in differentiating one material from another. Then a German physicist, Gustav Robert Kirchhoff (1824–1887), who frequently worked with Bunsen, had the key idea. Why study the colors by eye alone? Why not pass the colored light through a prism?

The two, Bunsen and Kirchhoff, developed the first "spectroscope" by allowing the light to pass through a narrow slit before passing through the prism. Each different wavelength of light was refracted by a characteristic amount and fell upon the screen in a particular place as the image of the slit in a particular color. If all the wavelengths of light were present, they would line up like soldiers in close-order drill, forming a continuous spectrum. Dense materials, heated to incandescence, produce such continuous spectra.

Vapors, however, produce only certain wavelengths of light. A mineral that, on heating, releases sodium vapors will produce, chiefly, light of two closely spaced wavelengths located in the yellow region of the spectrum.

By 1859, Kirchhoff, after studying numerous minerals, announced that every different type of atom produced its own pattern of spectral lines when heated to incandescent vapor. The different elements could then be identified by their spectral "fingerprints."

Of course, it was inevitable that eventually some mineral would be heated and would deliver a spectral line in a spot where no spectral line of any known element was located. The conclusion would be that a hitherto-unknown element had been detected.

On May 10, 1860, Kirchhoff announced that a certain mineral had produced a line in the blue region of the spectrum that was not identified with any lines produced by any known element. There was a new element present in the mineral that he named "cesium" (from a Latin word for "sky blue.") Once chemists knew cesium was there, they treated the mineral with usual chemical procedures and produced the element itself.

Within a year, Kirchhoff had discovered a red line that marked the existence of the element "rubidium" (from a Latin word for "red"). In the same way, other chemists discovered "indium" (from "indigo") and "thallium" (from a Greek word for "leaf green").

But what has this to do with the heavenly bodies?

Kirchhoff couldn't help but notice, in 1859, that the two closely spaced yellow lines produced by heated sodium vapor seemed to fall in the same place in the spectrum as did two dark lines of the Solar spectrum that were first recorded by Fraunhofer nearly half a century earlier. Fraunhofer had labeled those two dark lines as "D" and it seemed to Kirchhoff that the D-lines of the Solar spectrum had to have some connection with the yellow lines of glowing sodium.

Perhaps it was just an apparent connection. Perhaps the dark lines and the yellow lines were not *exactly* in the same place. Kirchhoff thought of a way of testing the matter. He would let sunlight pass through hot sodium vapor before passing through the slit of the spectroscope. The sodium vapor would supply yellow light that might just fit those wavelengths missing from sunlight, with the result that the Solar spectrum plus sodium light would erase those D-lines and leave a smooth, unsullied stretch of light.

It didn't happen! To Kirchhoff's surprise, the D-lines were still there. In fact, they were darker than they would be in the absence of the flowing sodium.

He experimented further and found this: A heated element emits certain wavelengths of light characteristic of itself. If, however, light from a source that forms a continuous spectrum is passed through vapor that is colder than itself, that vapor *absorbs* light of precisely the wavelength it would *emit* if it were hot enough.

Thus, sodium produces a characteristic double line of light in the yellow when it is heated. When light passes through sodium vapor from some source hotter than itself, the sodium vapor absorbs those same wavelengths and produces a double dark line where those wavelengths are now missing. When Kirchhoff allowed sunlight to pass through sodium vapor, the vapor did not supply yellow light, it absorbed it, making the dark lines darker. This effect of emitting when hotter and absorbing when cooler is called "Kirchhoff's law."

Now we can understand why there are dark lines in the Solar spectrum. Light is formed from heated layers of the Sun's blazing surface, and it produces a continuous spectrum because of the complexity of the dense structure. The continuous spectrum of sunlight passes through the vapors of the Solar atmosphere, which lies above the light-producing surface. The atmosphere, though hot by Earthly standards, is cooler than the surface and it absorbs some of the wavelengths of light, so that the spectrum when it reaches us and our instruments is crossed by thousands of dark lines indicating missing wavelengths.

From the dark D-lines in the Solar spectrum, we can deduce that sodium atoms are included in the Solar atmosphere; sodium atoms, moreover, that have the same properties in the Sun that they would have on the Earth.

Kirchhoff had, in this way, disproved Comte's categorical assertion of impossibility just two years after Comte's death, but not everyone was impressed. Kirchhoff had a banker, who, on learning that Kirchhoff had invented a way of detecting elements in the Sun, said (with uncommon stupidity—but, then, he was a banker), "Of what use is gold in the Sun if it cannot be brought down to Earth?"

When Kirchhoff was later awarded a medal and a prize in golden sovereigns from Great Britain for his work on spectroscopy, he deposited the money with his banker, saying, "Here is gold from the Sun."

Others began to study the Sun and other heavenly bodies by means of this new and very powerful technique. The Swedish physicist Anders Jonas Ångstrom (1814–1874) began to study the Solar spectrum in the finest detail and, in 1862, announced the presence of lines in the Solar system that were characteristic of hydrogen. This was important because, as it eventually turned out, hydrogen was the major component of the Sun, and of the Universe generally.

Ångstrom went on to discover lines characteristic of

other elements, and in 1868, he published a map of the spectrum locating with great care the wavelength of about a thousand lines.

What's more, he measured the wavelengths of those lines in units equal to a ten billionth of a meter. This unit was officially named the "Ångstrom unit" in 1905. To this day, it is still sometimes used for this purpose, but it is not part of the SI version of the metric system that is now official throughout the scientific world. Wavelengths of light should be measured in nanometers, where 1 nanometer is equal to 10 Ångstrom units.

The Sun was not the only luminous body studied. Ångstrom himself studied the spectrum of the aurora borealis in 1867.

More important still was the work of the British astronomer William Huggins (1824–1910). He studied the spectra of nebulae, of stars, of planets, of comets—of anything whose light he could make pass through a telescope, then a slit, then a prism, in order to produce a visible spectrum.

In 1863, he announced, from a study of spectral lines in the light of various stars, that the same elements that existed on Earth existed not only in the Sun, but in all the stars he studied. From that time on, scientists have been convinced that the eighty-one elements possessing one or more stable isotopes here on Earth are the only eighty-one elements possessing one or more stable isotopes anywhere in the Universe.

There have been times, to be sure, when it seemed that a new element was indicated by spectral analysis of light from a heavenly object, an element that, as far as anyone knew, did not exist on Earth.

The best-known case of this came in 1868, when a French astronomer, Pierre Jules César Janssen (1824–1907), was in India, studying light from the Solar corona, during a total eclipse of the Sun. He detected lines that he could not identify and he sent the data to a British astronomer, Joseph Norman Lockyer

(1836–1920), who was an expert on spectroscopy. Lockyer concurred with Janssen's suggestion that the lines represented an unknown element and he named it "helium," from the Greek word for "Sun."

For twenty-seven years, helium seemed to be what we might call an "Aristotelian element," one that occurred in heavenly bodies, but not on Earth. In 1895, however, a British chemist, William Ramsay (1852–1916), heard of a gas that had been obtained from a uranium mineral in the United States, one that had been identified as nitrogen, because it was odorless, colorless, and chemically inert (that is, reluctant to take part in chemical reactions). All this was characteristic of nitrogen.

Ramsay, however, had just participated in the discovery of argon, which was a gas that, like nitrogen, was colorless, odorless, and chemically inert, but was *not* nitrogen. (Argon was even more inert than nitrogen, for nitrogen would react with other substances in some cases, but argon never did.) The easiest way of distinguishing argon from nitrogen was to heat the gases strongly and study the spectral lines. The two elements might seem very similar in many ways, but their spectral fingerprints were completely different.

It seemed to Ramsay that the so-called nitrogen obtained from the uranium mineral might possibly be argon. Ramsay obtained a sample and studied the spectral lines it produced.

Right! It was definitely not nitrogen.

But it was not argon either. In fact—and here Ramsay must have been completely flabbergasted—the spectral lines were precisely those observed in sunlight during that eclipse of 1868. It was *helium*—and on Earth!

Since 1868, there have been other spectral lines detected that were not matched by anything on Earth. Other unknown elements were postulated and given names such as "coronium," "geocoronium," "nebulium," and so on. These were all false alarms. They didn't even represent new elements that were eventually found on

125

Earth as well. They turned out to be atoms of perfectly ordinary elements, such as iron, under conditions so extreme that the lines in question were never ordinarily produced on Earth.

If you're curious, we might summarize what we know of the composition of the Sun now. It is composed almost entirely of two elements, the two that happen to have atoms of the simplest structure: hydrogen and helium. The Sun is, in terms of mass, roughly three-quarters hydrogen and one-quarter helium.

All the other elements we know on Earth also exist in the Sun but in minor quantities. The elements other than hydrogen and helium make up perhaps 1.6 percent of the mass of the Sun. Of these minor elements, 50 percent of the mass is oxygen and 30 percent is carbon. All the other elements make up the remaining 20 percent of the 1.6 percent (that is, 0.3 percent of the mass of the Sun altogether).

This makes it look as though there *is* a basic difference in composition between the Sun and the Earth. Whereas the Sun is 98+ percent hydrogen and helium, 98+ percent of the Earth, according to most estimates, consists of six elements—iron, oxygen, silicon, magnesium, nickel, and sulfur. There is no overlap.

The difference, however, is not really fundamental. The entire Solar system formed out of a single cloud of dust and gas, with a composition much like that of the Sun. Objects that are very massive, like the Sun, have gravitational fields that can hold on to everything. Consequently the Sun has much the composition of the original cloud except for the conversion of some of its hydrogen to helium in the course of the nuclear fusion that has kept it blazing for 4.6 billion years.

If an object is moderately massive but cold (it is easier to hold on to light atoms in the cold), then again, the composition is that of the original cloud. Jupiter, Saturn, Uranus, and Neptune are probably almost entirely hydrogen and helium.

If an object is small, however, it lacks the gravitational pull required to hold on to hydrogen and helium, and it is made up of the minor elements in the cloud. Thus the satellites of the outer planets may be rich in icy materials (containing carbon, oxygen, and all the rest, and some hydrogen in combination with oxygen, forming frozen water).

Small objects that are also quite warm tend to lose all the volatile materials and are made up primarily of silicates (rocky substances) and metals. That is true of the bodies of the inner Solar system: Mars, Earth, Moon, Venus, and Mercury. Fortunately, the Earth is large enough to retain considerable water as well.

Part II
The Universe

9
The Importance of Pitch

When I was twenty-two, I married a beautiful damsel. (She was not my present dear wife, Janet, but that's another story.)

I was a little nervous about it. After all, I was neither handsome nor athletic nor wealthy nor sophisticated nor many other things that were likely to be attractive to women, and I was dreadfully afraid that the young woman would suddenly realize this.

I knew that I was intelligent, but I wasn't sure if that particular quality was very apparent (we had known each other only a few months) or, if it was, that it was of any importance. It seemed to me, then, that I must lose no chance to do something spectacular with it, something that might impress her.

Consequently, during our honeymoon at a mountain resort hotel, when it was announced one day that there would be a quiz contest that evening and that volunteers would be welcome, my hand went up at once.

I didn't think there was the chance of a snowball in

Hades that I would fail to win and I felt sure this would be bound to impress my new wife.

That night, I was third in line, and after the first two people had answered their questions, I stood up for my turn. At once the audience broke into spontaneous laughter. They hadn't laughed at the first two contestants, but I was very anxious, you see, and when I am anxious my face takes on a look that is even more intensely stupid than the one it wears in repose. So they laughed.

(My wife, who was in the audience, winced noticeably.)

The master of ceremonies then said, "Use the word 'pitch' in various sentences in such a way as to demonstrate five different meanings of the word."

The look of anxiety on my face grew more pronounced and the audience responded with wild hilarity. I paid no attention and merely collected my thoughts. When the laughter died down, I said, as loudly and as clearly as I could, "John pitched the pitch-covered ball as intensely as though he were fighting a pitched battle, while Mary, singing in a high-pitched voice, pitched a tent."

And then, in the dead silence that followed, I said (with a sly grin, I'm afraid), "One sentence does it."

Of course, I went on to win the contest and greatly impressed my wife. Interestingly enough, the affair won me considerable hostility from all the other guests. I gathered that there was a widespread feeling that I had no right to look so stupid without actually *being* stupid.

The reason I mention this now is that that little adventure of nearly half a century ago popped into my mind when I began to plan an essay in which I intended to describe how pitch (in the fourth sense used in that sentence of mine) told us a great deal about the size and the age of the Universe.

Let's start with sound.

Sound is produced when something vibrates. As it

moves in one direction, it compresses the air in the region into which it moves and rarefies it in the direction away from which it moves. Then it reverses direction in the course of its vibration, and the reverse happens. As the vibration continues, a large number of successive compressions are formed, with each one moving away from the sound source at the natural speed of molecular movement, given the temperature, pressure, and so on.

Sound, therefore, is a series of alternating compressions and rarefactions which, on striking the eardrum, causes it to vibrate after the fashion of the original vibration that set up the pattern. Through a complex series of physiological adaptations, the eardrum vibration is carried to the brain, which interprets what it receives as sound.

The compression-rarefaction alternation can be considered a wave form, and the distance from one compressed region to the next is the wavelength.

Ordinary objects, when set to vibrating, produce a vast complex of vibrations that, in turn, produce waves of a vast range of wavelengths that melt together in confused complexity to produce what our brain interprets as "noise."

There are objects, however, that will vibrate in relatively simple fashion and will produce sounds of a very small range of wavelengths. The brain interprets the result as a musical note, which is much more pleasant than noise is. By trial and error, primitive men discovered devices that made pleasant sounds, and it is combinations of these sounds that we call music.

The first scientific investigations of sound that we know of were carried out by the Greek philosopher Pythagoras (560-480 B.C.), who plucked strings of different length. He discovered that long strings vibrated more slowly than short strings did, and that long strings also produced deeper sounds than short strings did. In other words, differences in vibration (a physical fact) produced differences in pitch (a physiological interpretation).

133

Pitch also changes as the source of the sound moves toward you or away from you, but prior to the nineteenth century, this was not an easy thing to notice.

In the first place, the change in pitch is greater as the speed of the sound source increases, and in earlier centuries very few things moved quickly enough while producing enough sound to make the change in pitch noticeable.

In the second place, ordinary sound is noisy and the complexity of the wave form is such that it isn't easy to tell changes in pitch.

Perhaps you could tell the difference in pitch of a horn being played at a particular note if the hornist approached you on a galloping horse, passed you, and rode away from you. That, you must admit, is not a likely combination of circumstances.

In the 1840s, however, railroads were being built in Western Europe and in the United States. Trains of cars would chug along at a fair rate of speed, and in order to warn people to get out of the way, they had a one-note whistle which was sounded wildly when the train approached places where it was likely to encounter people.

That meant one thing became very noticeable. If you watched a train approaching while it was whistling its head off, you couldn't help but hear, as it passed you, that the whistle dropped suddenly and sharply in pitch.

To someone *on* the train, however, the whistle sounded a pitch that was lower than the pitch would seem to be to the person who was watching the train approaching, but higher than the pitch would seem to be to the person who was watching the train receding. What's more, to the person on the train, the pitch would remain constant.

Thus, suppose two people are standing near the railroad track and are a mile apart. The train is moving on the track between the two people. It has passed and is receding from the first person, and is approaching the second person. A person on the train hears the whistle at a certain pitch. The person who has already been passed hears it at a lower pitch than that, while the person who is still being approached hears it at a higher pitch than that. The three observers all report a different pitch at the same moment.

Why is that? Actually, the reason is very simple and I suspect that Pythagoras would have been able to figure it out if there had been trains with whistles in his day.

The reason was advanced in 1842 by the Austrian physicist Christian Johann Doppler (1803–1853). He reasoned in this fashion.

Suppose a train is at rest relative to an observer; that is, both the train and the observer are not moving at all, or the observer is on a moving train so that both are moving at precisely the same speed. In that case, the train's whistle sends out pulses of compression with regularity and you hear one, unchanging pitch.

But suppose the train is approaching you. The whistle sends out a wave of compression toward you. But the train is also approaching you so that the second wave of compression is released more closely to the first than would have been the case had the train been standing still. And the next wave is released more closely to the one before, and so on. All the waves of compression are closer together than they would have been if the train had been standing still. That means that the wavelength has been shortened and that, therefore, you perceive the pitch as higher than it would have been if the train had been standing still.

If the train is receding from you, matters are just the opposite. Each successive wave of compression is formed at a greater distance from the one before than would have been true if the train were standing still. The wavelength is longer and you perceive the pitch as being lower.

Doppler proceeded to work out a mathematical relationship that connected the pitch with the speed at which

135

the sound source was approaching or receding. This meant that from the change in pitch, a person could tell whether a train was approaching or receding, and at just what speed it was doing so.

The change in pitch with speed and direction of motion is therefore called the "Doppler effect."

In 1848, the French physicist Armand Hippolyte Louis Fizeau (1819–1896) pointed out that the Doppler effect is not confined to sound alone. Any wave form—in light in particular—shows a similar effect. This generalization is sometimes called the "Doppler-Fizeau effect" but Fizeau generally gets cheated because the lazy way out is to save two syllables by continuing to call it a Doppler effect even when it is applied to light.

Of course, light as we usually see it (from the Sun, from stars, from a kerosene lamp, from an incandescent bulb) is a complex set of waves of different wavelength; some wavelengths being so long or so short that we don't see them at all. Ordinary light, then, is analogous to what we call noise in connection with sound.

If a ray of light consisted of a single wavelength, that wavelength would shorten if the light source were approaching you, and lengthen if the light source were receding from you. Just as a particular wavelength of sound changes pitch as it lengthens or shortens, so a particular wavelength of light changes color as it lengthens or shortens.

Light of long wavelength is red. As the wavelength shortens, the color changes gradually through ranges of orange, yellow, green, blue, and violet in that order, the whole being referred to as the "light spectrum." Hence, if a wavelength of light lengthens because the source is receding from you, its color shifts toward the red end of the spectrum and this is called a "redshift." If the wavelength shortens because the source is approaching you, its color shifts toward the violet end of the spectrum.

This should be called a "violetshift," but scientists speak of it as a "blueshift" for reasons that pass my understanding.

But light sources that emit not one but a vast range of wavelengths would not ordinarily produce a noticeable shift at all. All the wavelengths would move toward the red or toward the violet. If they moved toward the red, some would fall off the red end, so to speak, and become invisible, while other wavelengths, ordinarily too short to be seen, would lengthen sufficiently to appear at the violet end of the spectrum. The same is true in reverse, if the wavelengths are all moving toward the violet end. In either case, what you actually see doesn't change much.

We can make an analogy. Imagine there's a long featureless rod and you can only see a small portion of it through a six-inch-wide slit. If the rod shifts in one direction or the other, you will continue to see only a small portion of it, and since it is featureless, you will not be able to tell how far it has shifted, or even in which direction.

If, on the other hand, there is some sort of mark on the rod, then you can at once tell the direction and extent of the shift, by noting the change in position of the marking.

As it happens, there *are* markings in light. In 1814, the German physicist Joseph von Fraunhofer (1787–1826) first noted that the Solar spectrum contains numbers of dark lines. These represent missing wavelengths of light in an otherwise continuous spectrum, because the Sun's atmosphere absorbs those wavelengths. Each dark line exists in a certain fixed place in the spectrum.

If the light source is approaching and all the wavelengths shift toward the violet, then the dark lines also shift toward the violet. If the light source is receding, then the dark lines shift toward the red. By observing the position of various dark lines, one can tell whether the light source is approaching or receding, and at what speed it is doing so.

What's more, the determination is distance-independent. It doesn't matter whether an object is nearby, or a few million miles away, or a few million light-years away. If a spectrum can be taken and the position of the dark lines noted, that is all you need.

But there's a difficulty. Sound moves quite slowly; only 0.331 kilometers per second (or 741 miles per hour). A train that is going at the moderate speed of 20 miles per hour is going at 2.7 percent the speed of sound and that is sufficient to produce a noticeable change in pitch.

Light, on the other hand, travels about 300,000 kilometers per second (186,000 miles per second) or just about a million times as fast as sound. If a light source were moving at 50 kilometers per second (31 miles per second), that would still be less than $\frac{1}{50}$ of 1 percent of the speed of light and that would produce only a very small shift in the dark lines of the spectrum.

It was not till 1868 that the British astronomer William Huggins (1824–1910) was able to study the spectrum of Sirius in sufficient detail to be able to note that there is a tiny redshift in its lines. Sirius is moving away from us at a brisk speed.

Over the next fifty years, the spectra of more and more stars were studied and the "radial motion" of each, whether toward us or away from us, was determined, and the speed of approach or recession estimated. The coming of photography was a crucial help in these studies since spectra could be taken through long-term exposures that couldn't possibly be seen by eye, and the position of the lines could be determined at leisure.

It turns out that some stars are approaching us, and some are receding from us. By analyzing certain regularities in these motions, scientists determined that the Galaxy is rotating about its center and they estimated the speed of rotation.

That's a pretty impressive result to arise out of something that began with the behavior of train whistles, but that is only the beginning.

In 1912, the American astronomer Vesto Melvin Slipher (1875–1969) managed to study the spectrum of what was then called the "Andromeda nebula." From the position of the dark lines in its spectrum, he found that it was approaching us at a speed of 200 kilometers per second. That did not seem out of line. Radial velocities of over 100 kilometers per second were unusually high, but not distressingly so.

(We know today that part of the velocity is not to be attributed to a true approach of the Andromeda. Actually, the Andromeda is a distant galaxy, something not known in 1912. The rotation of our Galaxy is, at the moment, carrying us toward Andromeda. If the effect of this rotation is allowed for and the motion of Andromeda is measured relative to the center of the Galaxy, it is found to be approaching us at a velocity of only about 50 kilometers per second.)

By 1917, things began to seem really puzzling, however. Slipher had gone on to measure the radial velocity of a total of fifteen nebulas that resembled Andromeda in having a spiral form. On the basis of sheer chance, one might have expected that half of them would be approaching and half receding. Instead, the Andromeda and one other galaxy were approaching, while the other thirteen were receding.

Actually, this was more puzzling than Slipher knew at the time. All the spiral nebulae he studied were actually distant galaxies. The two that were approaching were relatively close to us and were part of the "Local Group"—a cluster of galaxies, including our own Galaxy and the Andromeda, all held together by gravitational forces, and all revolving about the center of gravity of the entire group. Each one would be approaching us at one epoch in the history of the Universe and receding from us at another.

The other thirteen, which are *not* part of the Local Group, are all receding from us, which is a peculiar coincidence. They might move, in response to gravity, through still larger orbits and also might be receding at some times and approaching at others. Slipher might just have happened to measure this particular bunch of thirteen galaxies at a time when each one was in the recession phase of its orbit. That is really unlikely, but not entirely impossible. You *might* toss a coin thirteen times and end up with thirteen tails just as a matter of chance.

What was more disturbing was the radial velocity of the thirteen. Their recession was at an average velocity of 640 kilometers per second. While 200 kilometers per second could be swallowed, 640 kilometers per second was very difficult to accept. It was much greater than the radial velocities of the stars about us.

Slipher continued to measure radial velocities of more and more nebulae and found that, without exception, they all showed redshifts and were therefore receding from us.

In the 1920s, the nebulae were finally identified as distant galaxies, and in part, that eased the situation. The galaxies were objects that were completely different from the ordinary stars about us, and it might well be normal for them to move much more quickly relative to other galaxies than for a star within a particular galaxy to be moving relative to other stars within that same galaxy.

But there remained the sticking point: Why should all those radial velocities be recessions? Shouldn't there be at least *one* galaxy outside the Local Group that was approaching us? You might think there should be—but there wasn't.

Things just continued to get worse. The American

140

astronomer Milton LaSalle Humason (1891–1972) carried on Slipher's work. He took photographic exposures that lasted for days, so that spectra could be recorded for fainter and fainter galaxies. When he did so, he discovered velocities of recession that made earlier observations seem piddling in comparison. In 1928, he photographed a galaxy that was receding from us at 3,800 kilometers per second, which is 1.25 percent of the speed of light. By 1936, he was clocking velocities of 40,000 kilometers per second, 13 percent that of light. And *still* only recessions showed up. No approaches.

Why should this universal recession take place? Could it be that the red shift did not signify a recession? Could it be produced by other factors?

For instance, light traveling long distances through the thin gases occupying the vast spaces between the galaxies might simply redden, as sunlight reddens when the Sun is at the horizon and its light must pass through an unusual thickness of atmosphere.

Undoubtedly, but in that case, the reddening is just the result of the scattering of shortwave light. The position of the dark lines is not affected.

Might it be then that light when it travels long distances simply loses energy? If it does, it would naturally shift toward the red, since the longer the wavelength, the smaller the energy content of the light. In that case, there would be no reason to suppose the galaxies were receding at all. The light was merely getting "tired."

The trouble is that physicists don't know of any process that would allow light to lose energy merely by passing through space. Besides, if it did lose energy in that fashion, then light, traveling even fairly short distances, ought to get very slightly tired, and lose a tiny bit of energy. This should be detectable in studying objects within our own Galaxy and perhaps even within our own Solar system—but the effect has never been detected.

Again, the reverse might be true. Instead of the galaxies being very distant objects that are moving slowly,

141

they might indeed be moving quickly, but they might be relatively nearby and they might be small objects and not galaxies at all. Perhaps they are objects that have been shot out of the few galaxies that *do* exist, perhaps even out of our own. They might be moving at very high velocities but that wouldn't mean they were far away—they had merely been shot out with huge energies.

This has been made to seem more plausible in recent years when it has been discovered that some galaxies are very active and have centers where enormous energies are released. Perhaps these can explode and eject matter.

If they do, however, it seems to pass the bounds of credibility that all the expelled masses are moving away from us. Perhaps that is because those masses that were moving toward us have passed us and are now moving away from us. But that doesn't help. Surely, there must be at least *one* bit of matter that was shot out in our direction and has not yet passed us; *one* object that seems to be a galaxy and is approaching us.

But there isn't, outside our own Local Group. Not one.

Astronomers have simply failed to account for the redshift in any plausible way except by means of the good old Doppler effect. The galaxies are simply receding from us, and at incredible velocities.

The American astronomer Edwin Powell Hubble (1889–1953), who was working along with Humason, tried to estimate the distances of various galaxies. The nearest ones had certain variable stars called "Cepheids" that could be made out individually. From their period of variation, it was possible to calculate their luminosity—how much light they emitted. From that and from their apparent brightness in the sky, one could calculate their distance, which would also be the distance of the galaxy that contained them.

If a galaxy was so far away that no Cepheid variables were bright enough to be made out, some supergiant stars could be. Assuming that such supergiant stars would be as luminous as those in nearer galaxies, the distance of the distant galaxies could be worked out.

If a galaxy was so far away that no stars at all could be made out, Hubble went by the luminosity of the entire galaxy, assuming that the dimmer it was, the farther away it was.

Having estimated the distances of a number of galaxies, Hubble checked each with the velocity of recession that had been calculated for them. He found that, on the whole, the velocity of recession of a particular galaxy was directly proportional to its distance from us.

This means that if galaxy A is x times the distance of galaxy B, then galaxy A is receding at x times the velocity of galaxy A.

This is called "Hubble's Law."

The most astonishing feature of Hubble's Law—the notion that all the galaxies are moving away from us, the farther the faster—prompts a simple question: "Why from *us*?"

Back in 1935, the science fiction writer Edmond Hamilton (1904–1977) published a story entitled "The Accursed Galaxy," which gives a very interesting explanation.

Hamilton suggests that, originally, all the galaxies were comparatively close together and were very nearly at rest relative to each other, except for orbital velocities enforced by gravitational constraints.

But then, in one particular galaxy (our own, of course) life somehow developed. This was a serious galactic disease that looked as though it would rapidly spread through the Milky Way, infecting every region and passing on to any other galaxies that were too close.

All the other galaxies, in wild panic, have been fleeing from us ever since, and those which managed to achieve higher speeds managed to make it farther away in the time since the infection appeared.

This is a delightfully ingenious notion that one might

almost feel ought to be true because it is so pretty, but of course, it is just fantasy. It attributes a *purpose* to the recession and that's outside the rules of the game of science. Things should happen only in blind obedience to the laws of nature.

Let us then carry on with the problem of the receding galaxies in the next essay.

10
Long Ago and Far Away

Several months ago, I walked out at the close of a banquet and found it to be raining briskly. It was plain there would be no taxis, so two other banqueteers and I made our way to the nearest subway entrance, got on a subway train, and trundled northward.

As it happened, my stop came first. I bade my friends farewell and got off the train. The next day I found out what had happened to them after I got off.

Three youngsters walked up to where my friends were sitting, and towered over them in what seemed to them to be a threatening manner. My friends were well aware of the violence that sometimes takes place in the subway and they were naturally apprehensive.

One of the youths said something in a low voice and one of my friends, plucking up his courage, said, "I'm sorry, young man, I didn't hear you. Would you repeat it, please?"

Whereupon the young man, in a louder voice, said,

"What I asked was: Was that Isaac Asimov that just got off the train?"

In a flash, the youngsters had changed from three threatening hoodlums into three concerned fans of culture with impeccable taste, and my friends answered cheerily that indeed it was, and all was wine and roses thereafter.

I don't know if those intelligent young men on the subway ever read my science essays, but if they do, this one is dedicated to them.

In the previous essay, I talked about the Doppler effect and the way in which it was used to show that the distant galaxies were all, without exception, receding from us, and that the farther off a galaxy was, the faster it was receding.

That gives our Galaxy a peculiar distinction, it would seem. It is the one from which all others recede, the farther the faster. That doesn't seem right, somehow. Why should our Galaxy, among all the billions, have that distinction?

Well, back in 1916, Albert Einstein (1879–1955) devised a set of "field equations" that describe the characteristics of the Universe as a whole. Einstein assumed that the Universe as a whole is static and doesn't perceptibly show any progressive change with time—and after all, there wasn't anything in our observation of the Universe up to that time that indicated anything different. That is, some objects in the Universe might be approaching us, some might be receding, some might be going in this direction and some in that, but all these changes would tend to cancel and the overall appearance of the Universe would remain the same.

Einstein's equations didn't quite demonstrate this, so Einstein arbitrarily added what he called a "cosmological constant," and gave it the value required to allow his

146

equations to represent a static Universe. (He later called this his "greatest mistake.")

The next year, however, the Dutch astronomer Willem de Sitter (1872–1934) pointed out that the field equations, without the cosmological constant, represent an expanding Universe, one that is growing larger at a constant rate. That expansion seemed purely theoretical, however, for there was at that time no indication that the Universe is indeed expanding.

However, once Hubble showed that the distant galaxies are all receding from us, that represented observational evidence that de Sitter had made a valuable point. The Universe *is* expanding, and all the galaxies (or clusters of galaxies) are moving away from each other. That is what makes the galaxies all appear to be receding from us, and it does not require us to be living in a special galaxy. If the Universe is expanding, then if we observe the galaxies from *any* galaxy, all the other galaxies will seem to be receding from each other, the farther the faster.

Einstein's field equations, in other words, describe the Universe as it is, and no cosmological constant is necessary.

If the Universe is expanding and we consider the future, it might well be that the Universe will continue to expand forever. There is, presumably, no limit to the space into which it can expand.

On the other hand, if we consider the past, we see that the Universe must have been smaller and smaller as we move farther into the past. That, in turn, means the past of the Universe, unlike the future, cannot last forever. At some time in the finite past, the Universe must have been so small that all its mass and energy were compressed into a small ball.

This was first pointed out, in 1927, by the Belgian

147

astronomer Georges Edward Lemaitre (1894–1966). He called that small blob of mass the "cosmic egg." The cosmic egg apparently exploded to form the Universe as it now exists, and the Russian-American physicist George Gamow (1904–1968) called the explosion the "Big Bang."

The Big Bang theory is now generally accepted by astronomers. There is some argument where the original cosmic egg came from, how it formed, how large it was, by what initial stages it developed into the Universe we now have, and so on. Such things are not the question of interest in this essay, however. Instead, let's ask, simply: *When* did the Big Bang take place? How long ago?

The way to get the answer is to determine how rapidly the Universe is expanding right now. Then, if the rate of expansion doesn't change with time, the moment of origin may be directly determined from that, and very simply, too.

Hubble's measurements, back in 1929, made it seem that the rate of recession is indeed constant, and that the Universe is expanding so rapidly that, looking back in time, the cosmic egg must have existed something like 2 billion years ago.

That this had to be wrong was indeed quite obvious, for geologists were certain that the Earth itself is older than that, and the Earth could scarcely be older than the Universe of which it is part. From the decay of radioactive substances in the Earth's crust, in the Moon, and in meteorites, we now know that the Earth, and the Solar system generally, are 4.5 billion years old, so the Universe must be at least that old and possibly considerably older.

Fortunately, the yardstick used to determine the distance of the nearer galaxies, which, in turn, served as the basis for determining the distance of all farther galaxies, turned out to be a far more complex matter than had been supposed. When the yardstick was revised, it was clear that the Universe is considerably larger than had been thought, and must, therefore, have been ex-

148

panding for a longer time. It followed that its age is correspondingly larger.

The time of the Big Bang must then have been at least 10 billion years in the past and, very likely, more. In fact, the most common figure now used is 15 billion years in the past.

If the Universe is billions of years old, then we should see objects in the sky that are billions of light-years away, and if so, we shall be seeing far into the past. After all, if a star is 10 light-years away, then it takes the light from that star 10 years to reach us and we see it as it was 10 years ago. If a galaxy is 10 million light-years away, the light takes 10 million years to reach us and we see it as it was 10 million years ago.

Whatever is far away, in other words, is long ago, and the farther away it is, the longer ago it is. If we can see *very* distant objects, then, we will be seeing the early days of the Universe. After all, if we see something that is 15 billion light-years away, we will see it as it was 15 billion years ago, which would show it to us at the very beginning of the Universe.

Yet the farther an object is, the dimmer it is, and the more difficult it is to make it out. Astronomers at their telescopes in 1960 could not have been very hopeful that we would have much chance of seeing the early days of the Universe.

By 1960, for instance, the most distant galaxies they could see were galaxies that, judging by their redshifts and the new yardsticks, were perhaps 800 million light-years away. That means they were seeing only 800 million years into the past, and if the Universe is 15 billion years old, they were seeing only about one twentieth of the way back to the beginning.

How on Earth could they do better than that?

By 1960, however, astronomers had radio telescopes, which detected radio waves rather than light. There didn't seem to be any reason, at first, to suppose that radio waves would tell us more about the Universe than light waves would. For instance, the Sun emits radio waves, but these can scarcely be as useful as light is in telling us about the heat, the chemical composition, and other characteristics of the Sun.

Moreover, whereas light comes from a myriad of stars in this and other galaxies, the number of radio wave sources is far fewer. Distant stars which deliver enough light to seem quite bright even to the unaided eyes don't deliver detectable amounts of radio waves. The only reason we detect radio waves from our Sun is not because it is an unusual star, but because it is so near to us.

There are, however, radio sources that seem to originate from so restricted a portion of the sky that they were called "radio stars" in the old days. It was not necessarily assumed, on the other hand, that the radio stars are actual *stars*. They might originate in galaxies that are so distant they don't take up very much apparent space in the sky. An entire galaxy might, after all, give off enough radio waves to be detected, even if individual stars don't.

Some radio stars are *so* compact, however, that the thought arose that they might, after all, *be* stars. Among these very compact radio sources are several known as 3C48, 3C147, 3C196, 3C273, and 3C286. [The "3C" stands for *Third Cambridge Catalogue of Radio Stars*, which was compiled by an English astronomer, Martin Ryle (1918–1984) and his colleagues.]

In 1960, the areas containing these compact radio sources were combed by the American astronomer Allan Rex Sandage (b. 1926), using the 200-inch Palomar telescope. In each case, a star did indeed seem to be the source of the radio waves, and they seemed undistinguished dim stars of our own Galaxy.

There were, however, some disturbing notes. Why should these few stars emit radio waves of detectable intensity when other nearer and brighter stars do not? Then, too, when these radio stars were examined very closely, faint nebulosities were associated with some of them. As for 3C273, the brightest of the bunch, there were signs of a tiny jet of matter emerging from it.

These radio stars, though they looked like stars, seemed to be something different. They were called "quasi-stellar radio sources," where "quasi-stellar" means "star-resembling." As the term became more and more important to astronomers, it became too inconvenient a mouthful and was shortened to the first and last syllables of quasi-stellar—"quasar."

Clearly, the quasars were interesting enough to warrant investigation with the full battery of astronomic techniques, which meant that their spectra were needed. Getting spectra of dim objects is not easy, but the American astronomer Jesse Leonard Greenstein (b. 1909) and his Dutch-American colleague Maarten Schmidt (b. 1929) labored at the task and obtained the spectra.

When they did, it didn't help. They found the quasars had strange spectral lines that couldn't be identified. What's more, the spectral lines of one quasar were different from all the others, and all of them were unidentifiable—which at least added to the feeling of strangeness about the objects.

In 1963, Schmidt noted that of the six lines present in the spectrum of 3C273, four were spaced in such a way as to resemble a series of hydrogen lines—except that no such series ought to exist in the place in which they were found. What, though, if those lines were located elsewhere but were found where they were, only because they had been displaced toward the red end of the spectrum? If so, it had to be a large displacement, one that indicated a recession at a velocity of 40,000 kilometers per second, or well over an eighth the speed of light.

This seemed unbelievable, and yet if such a displacement existed, the other two lines could also be identified; one represented doubly charged oxygen, and the other doubly charged magnesium.

The quasars, it would seem, then, are not stars of our

own galaxy at all. They are objects where even the nearest is at least a billion light-years away. (They would not have been discovered except for radio telescopes, and their significance would not have been understood save for the redshift, hence "the importance of pitch," which I discussed in the previous essay.)

Obviously they must be extremely bright to be made out even in our best telescopes at such a distance. They must, in fact, be a hundred times as luminous as a galaxy such as ours. Naturally, there must be something extremely unusual going on inside them to produce all that light and all those radio waves.

What's more, it was found, by 1963, that quasars vary in brightness, sometimes surprisingly quickly, and the variation comes in both the light and the radio waves. The variations are large and are noticeable over a period of a year or so.

This means that quasars must be very small in size. Small variations might result from brightenings and dimmings in restricted regions of an object, but large variations must involve the object as a whole. If the object is involved as a whole, then some effect must make itself felt across the full width of the body within the time of variation. Since no effect can travel faster than light, a marked variation in the space of a year indicates that the quasar cannot be more than a light-year in width. Some quasars seem to be even smaller than that.

Such combinations of great luminosity and tiny volume are puzzling indeed.

The possible answer to this came from a study of galaxies in general by means of radio astronomy.

If we look at galaxies by ordinary light, they seem beautiful, but quiet. The centers are brighter than the outskirts because stars are more densely distributed in the centers than in the outskirts.

Radio astronomy shows us, however, that from the

very core of many galaxies there is a steady outflow of huge amounts of energy. This is true, to some extent, even of our own Galaxy. We can't see the core of our Galaxy by the light it emits because dust clouds in the way block our vision. Radio waves go right through the clouds, however, and our radio telescopes tell us that in a very small volume right at the core, there is a veritable flood of radio waves.

Increasingly, astronomers are of the opinion that there are large black holes at the center of many (and perhaps of *all*) galaxies, and that the energy is produced by the swallowing up of matter, even of entire stars, by the black holes at the center.

Where, for some reason, the central black hole is particularly massive and particularly active, then particularly large floods of radio waves emerge—a great deal more than are produced by our own quiet and respectable Galaxy. Where the black hole is raising the roof, we have what is called an "active galaxy."

Naturally, the core of an active galaxy ought also to give off an extraordinarily high intensity of light, and such a core would gleam much more brightly than the rest of the galaxy.

Back in 1943, an American astronomer, Carl Seyfert, observed an odd galaxy with a very bright and very small nucleus. Other galaxies of the sort have been discovered since and they are now termed "Seyfert galaxies." Some astronomers feel that these may represent as much as 1 percent of all galaxies.

It may be, then, that quasars are very large or very extreme Seyfert galaxies, ones that are so distant that only the incredibly bright center can be seen and that makes the whole object look like a star. Indeed, in recent photographs, the dim nebulosity around the quasars has shown up more clearly and it seems very likely that we are dealing with bright Seyfert galaxies.

The quasars are all very far away. Not a single one is less than 1 billion light-years away. Generally, they are much farther. It is tempting to suppose that quasars are the products of the energetic youth of our Universe and that their prodigal expenditure of energy soon exhausts them. As the Universe grew older, more and more of the quasars dimmed and settled down, and fewer and fewer new ones were spawned, until in the last billion years there were none that existed, and perhaps none will exist in the future either.

Quasars can be seen at very great distances, particularly with modern techniques for detection. However, there must be a limit beyond which they cannot be seen.

Suppose the Big Bang took place 15 billion years ago. There were probably stages in the youth of the Universe in which energy dominated space, and space, considering the thick mix of photons that was present, was not transparent. As the Universe expanded and cooled, the energy condensed into matter, space grew transparent, and eventually the galaxies, including the quasars, were formed.

If we gaze through our telescopes, optical or radio or anything else, we would eventually penetrate to places so far away, and therefore so long ago, as to see nothing but that opaque haze that marked the Universe before the stars and galaxies were formed. We would see that haze in every direction and that would mark the "end of the Universe."

But beyond even the haze there must be the Big Bang itself, and we ought to detect the radiation from that. You might think it ought to be seen as an unbelievably brilliant burst of radiation, but it is so far away that the enormous redshift puts it all into the radio wave region.

In 1949, Gamow suggested that the radio wave radiation from the Big Bang ought to be detectable everywhere in the sky at the same intensity. The American physicist Robert Henry Dicke (b. 1916) took up this notion and carried it further.

With Dicke's help, the German-American physicist Arno Allan Penzias (b. 1933) and the American radio astronomer Robert Woodrow Wilson (b. 1936) actually detected this radio wave background in the sky in 1964. This was the strongest indication yet that the Big Bang did indeed take place.

Given the cutoff region of the Big Bang plus the opaque fog of radiation around it, what are the farthest things we can see short of that limit?

In 1965, Maarten Schmidt discovered that the quasar 3C9 has so huge a redshift that it must be 10.5 billion light-years away and receding from us at a velocity of 240,000 kilometers per second. This is 80 percent the speed of light, and it didn't seem likely there would be anything much farther away than that.

In 1973, the redshift of quasar OQ172 was determined, and it was found to be 11.5 billion light-years away. (This means, by the way, that the Universe *can't* be only 10 billion years old, unless the redshift is interpreted as signifying something other than distance according to the Hubble formula—and some astronomers suspect this might be so. They are, however, a small, if sometimes vocal, minority.)

Some 1,500 or more quasars were discovered after 1973 and none clearly beat the OQ172 record.

Astronomers have now gotten into the habit of referring to redshifts according to the percentage of the shift as compared to the original position of the line if the emitting object were at rest. The percentage is divided by a hundred, so that if a line is shifted by 100 percent, it has a redshift of 1; if it is shifted by 200 percent, it has a redshift of 2; and so on.

The redshift of OQ172 is about 3, but then, in 1987, things broke loose. Making use of new techniques for scanning the sky, places near the south Galactic pole were chosen. These are as far away from the Milky Way as possible so that there is no dust obscuration and devices can penetrate into very deep space. In this way,

155

fourteen quasars were found with redshifts greater than 3 and two had redshifts greater than 4. In fact, redshifts as high as 4.43 are now known.

Astronomers don't know whether still higher redshifts may be found. If so, the astronomers are going to find themselves in an uncomfortable situation, for the best theories of galaxy formation indicate that they first formed at a time equivalent to a redshift of 5. If we go beyond that, astronomers will have to work up new theories of galaxy formation.

In fact, they may be in trouble, in any case, because of another far-off discovery, not involving quasars. After all, quasars are very special kinds of galaxies, and may not be representative of the whole. How far off can we see ordinary galaxies?

The problem, there, is that ordinary galaxies are considerably dimmer than quasars and are much harder to see. Nevertheless, there are new techniques for making out extremely dim objects and we can see things that were hopelessly beyond our reach only a few years ago.

The American astronomer J. Anthony Tyson led a group that made use of a large radio telescope in Chile plus something called a "charge-coupled device" to record the images.

They chose twelve regions of the sky, each about 3 minutes of arc by 5 minutes of arc, so that each single region is about $\frac{1}{200}$the size of the full Moon, and all twelve together are $\frac{1}{17}$the size of the full Moon. These twelve portions of the sky are far away from the Milky Way, and they lack any bright stars or galaxies. They seemed, essentially, to be empty space.

The new techniques, however, found up to a thousand fuzzy objects within each one of these samples of "empty space." In all twelve samples there are about 25,000 objects.

These fuzzy objects are not point sources as stars would be and they are not bright enough to be quasars. The feeling is that they are ordinary galaxies, or at least "primeval galaxies." It seems possible that since these primeval galaxies swarm in all twelve areas chosen, they swarm everywhere, and there may be 20 billion of them in the sky altogether.

Those that are present are at the "confusion limit." That is, if there were any more (or if we could see still dimmer ones), they would overlap and not be seen as individual objects at all.

The redshifts of the brighter of these primeval galaxies range between 0.7 and 3, which puts them at distances of from 7 to 11.4 billion light-years. It may be that some have redshifts of 4 or more and date back to a time only a billion years or so after the Big Bang.

Why are they packed together so tightly? Well, if we look in any direction at all for a distance of, say, 10 billion light-years, we're looking at a Universe that may have only 4 percent the volume of the present Universe. The galaxies in that early Universe would be separated, on the average, by only $\frac{1}{25}$ the distances that separate them today, so naturally they're packed closely together.

As we look outward in different directions, we seem to be seeing a vast shell of primeval galaxies surrounding our Universe, but we're actually looking at the same small Universe from different angles.

Yet if distant primeval galaxies really exist—if they are not some incredible mistake in the instrumentation or the interpretation—they tell us that galaxies began to form only a billion years after the Big Bang and continued to form at a gradual rate for some 5 or 6 billion years.

Since current theories of the development of the early Universe tend to picture the galaxies as forming somewhat later, to begin with, and doing so in a great burst, it would seem that our view of the Universe must be changed. This could be very exciting for we should end up with a much better and more satisfying picture of the Universe, and one which is closer to reality.

11
The Cosmic Lens

There came a time in my life when it dawned on me that other people were thinking of me as a sort of celebrity, and it made me uneasy.

I was brought up by poor but honest parents, you see—very honest and very poor. I've always thought of myself as belonging to the same category, but I couldn't help thinking that honesty was the more important of the two so I strove to correct the other characteristic. Yet I couldn't help feeling that as the money piled up, the opportunities for dishonesty would grow—and what if I was lured into that vice.

So I tried to ignore the fact that I was getting unreasonably well off and continued to live simply and frugally (as far as was consistent with reasonable comfort, especially for the children). That way I could stay poor but honest.

The game wasn't going to work, however, if I was going to be treated as a celebrity, so I grew very determined not to "go Hollywood"; not to expect the privileges of rank; not to demand or even accept special treatment.

Let me tell you, it's hard. And there *are* times when I can't resist temptation. I will give you an instance.

Some time ago, I was waiting at my usual corner for a taxi, and I was in a reasonable hurry and no unoccupied taxis were in sight. Naturally, I began to grow unhappy.

A taxi skidded to a halt in front of me, but it contained a very noticeable fare in the back seat, so I paid no attention to it, but continued to look glumly in the direction of oncoming traffic. The front window cranked down, however, and the cabbie leaned out. He had obviously recognized me.

"Dr. Asimov," he said, "I'm a fan of yours and I'd pick you up, but I have a fare."

"Thank you. I appreciate your thoughtfulness," I said.

The taxi was about to move on, but the rear window cranked down and the fare leaned out and said, "Dr. Asimov, I'm a fan of yours, too. Get in!"

What I should have said at that point was, "No, thank you. I want no special treatment. I'll wait for an empty taxi like anyone else."

But I didn't. I got in. I tried to make up for it by insisting on paying for both of us, but what it amounts to is that I took advantage of my position and accepted privilege. It bothers me.

And yet—Every month I take advantage of my position and accept privilege. Do you suppose that anyone else could talk our Noble Editor into letting him do a monthly science column? Of course not. I'm the only one—because I'm old and venerated and have been doing it for years. Ought I to say, "Mr. Ferman, I want no special treatment. Throw this spot open to anyone, and take the best."

No *sir*, I won't do it. Not in a million years. And it doesn't bother me either, so let's get on with it.

* * *

In 1916, Albert Einstein (1879–1955) advanced his general theory of relativity and pointed out that, if the theory was correct, light ought to curve in its path in a gravitational field. The amount of curvature would be immeasurably small under ordinary circumstances (on Earth, for instance), but circumstances might be extraordinary.

Suppose the light rays from a star passed very near the rim of the Sun on their way to Earth. In that case, the curvature (toward the Sun) might be great enough to make the star seem slightly farther from the Sun than it ought to be, because the eye would follow the ray back, so to speak, in a straight line, not allowing for the curvature.

Naturally, a star that would be so close to the Sun that its light would just skim the solar disk would not be seen in the Sun's glare—*unless* there happened to be a total eclipse at the time.

But in 1916, Europe was in the very midst of World War I and eclipse expeditions were out of the question. Still, the British astronomer Arthur S. Eddington (1882–1944) obtained a copy of Einstein's paper and began making plans.

The war came to an end on November 11, 1918, and that left time to arrange eclipse expeditions for May 29, 1919, the day an eclipse was expected when the Sun happened to be in the midst of an array of bright stars.

Eddington organized two expeditions, one to northern Brazil and one to an island off the coast of West Africa. The positions of the bright stars near the Sun were measured relative to each other, and this was compared with those same relative positions a half year later when the Sun was in the opposite section of the sky. The results were borderline but were taken to support Einstein. (Since then, of course, similar experiments, carried out far more precisely, have left no doubt that Einstein was correct.)

The notion of the curvature of light by gravitational

fields gave rise to some interesting speculation. The Sun is so large that a ray of starlight can pass to one side or another and be bent. But what if the ray of light from a distant star passed by the rim of another star but one nearer to itself? The ray of light might be thick enough so that part of it would pass to one side of the nearer star, and part to the other side.

Both parts of the ray would be bent inward toward the nearer star. Therefore, the eye of an observer on Earth would follow one part of the ray in a straight line in one direction and the other part in a straight line in another direction. Instead of seeing one distant star beyond the nearer star, we would see *two*, one on one side of the nearer star and one on the other.

This was first suggested in a casual way in 1924.

In 1936, Einstein himself took up the matter and subjected it to careful mathematical analysis. The nearer star by splitting a ray of distant starlight in two and bringing the two to a focus, more or less, on Earth, was acting in a fashion analogous to that of a glass lens. Einstein therefore called the phenomenon a "gravitational lens."

He showed that, indeed, the distant star would be seen as two stars. If the distant star were a tiny bit to one side of the center of the nearer "lensing" star, more of its light would skim by that side of the star, less by the other side. Therefore, of the pair of stars that would be seen, one would be brighter than the other, but both would have precisely the same spectrum in all respects, since observers would really be seeing the same star.

If, however, the more distant star was *precisely* behind the nearer star, the light from the distant star would pass by the rim of the nearer star on every side and we would see the nearer star surrounded by a tiny ring of light that would be the lens distortion of the more distant star. That would be an "Einstein ring."

This was fascinating theory, but there seemed (to Einstein in 1936, and to other astronomers of the time) no chance at all of finding examples of gravitational lenses in the sky.

In the first place, despite the vast number of telescopically visible stars in the sky, the chances of finding one star exactly, or almost exactly, behind another is so excessively tiny that looking for an example was quite likely to be labor wasted.

Besides, even the gravitational intensity at a stellar surface is not very large in terms of star light. The light is curved so slightly that the two rays from opposite sides of the star would have to travel an enormous number of light-years before coming close enough to focus. In other words, any gravitation lens in the form of a star would have to be very far away from the Earth, and the more distant star whose image is being distorted would have to be much farther away still.

In fact, the distorted image would be so far away that it was likely to be too dim to be seen at all and there could certainly be no useful spectrum to be gotten from it. Therefore, one couldn't judge whether two stars that seemed very close in the sky might have the same spectrum and therefore possibly be two images of a single star, or have different spectra and be two distinct and unrelated stars.

Of course, some stars have more intense gravitational fields than others do. In 1915, the American astronomer Walter Sydney Adams (1876–1956) had discovered that Sirius's companion star, Sirius B, was what we now call a "white dwarf." It packed all the mass of a star like the Sun into a condensed body smaller than the Earth so that the gravitational field near its surface was tens of thousands that of the surface of an ordinary star. (He proved this by demonstrating, in 1924, that the light from Sirius B had a redshift produced by its intense gravitational field, as Einstein's general relativity had predicted.)

For that reason, if it was a white dwarf that acted as a lensing star, a distant star's rays would be curved more sharply and the focal length would be shorter. The distorted image might then be nearer to us and, therefore, bright enough to be visible to us. There are so few white dwarfs as compared with ordinary stars, however, that the shorter focal length is more than made up for by the greater unlikelihood of a gravitational-lens situation at all.

If, then, all we had to deal with were stars, Einstein's theoretical investigations would represent the end of the story. The Universe is *not* just a matter of stars, however, as was quite evident even in 1936.

In the 1920s, it was established that certain "nebulae" (small fuzzy objects) were actually independent galaxies located far outside our own. There were thousands and millions of them spread out over millions, tens of millions, and hundreds of millions of light-years.

It follows, then, that if we're talking galactic lenses, we oughtn't to talk just about stars-behind-stars. But hindsight is cheap. Right now, I can see the inevitability of going beyond stars with half an eye and a hundredth of a brain, but in 1936, Einstein himself, lacking the benefit of hindsight, didn't see it.

It was in 1937, that the Swiss astronomer Fritz Zwicky (1898–1974), who was interested in the distant galaxies, did see it. He pointed out that as the Universe was revealing itself to be richer and richer in galaxies, it might be only a matter of time before one galaxy would be discovered that was exactly in front of another, more distant galaxy; or perhaps one cluster of galaxies would be exactly in front of another, more distant cluster; and then gravitational-lens effects might be noted.

Let's see what the advantages of the galaxy-galaxy gravitational lens would be as compared with the star-star situation.

In the first place, stars are essentially points of light, so that the chance of one point of light being positioned

163

right behind another, as we view them, is very slight. Galaxies, on the other hand, are extended bodies; small, of course, but far from points. It is possible then that even if galaxies are not one behind the other, center to center, there might nevertheless be a partial overlap and that may be enough to produce a gravitational-lens effect. A galaxy overlap is therefore much more likely than a star overlap.

Second, since galaxies contain anywhere from a billion to a trillion stars, they can be seen, and their spectra studied, at hundreds of times the distance of even the brightest individual stars of the ordinary kind. The more distant an object is, the greater the number of other objects likely to be lying more or less between it and ourselves, and the greater the chance that one of those intermediate objects may be close enough to the traveling ray of light from the distant object to produce a lens effect.

Third, the more distant an object is, the more likely it is that the long focal length of a galactic lens may come sufficiently to focus at Earth to allow us to see the distortions that have been produced.

Are there disadvantages? Yes. In the star-star situation, where you are dealing with two point sources of light, the lens effect is a simple one and you will get either two stars or (very unlikely) an Einstein ring.

In the galaxy-galaxy situation, with two extended sources which are not evenly bright from point to point, the lens effect is much more complicated. You can get three images or five. These can be asymmetrically placed and show tortured shapes that may not seem to be lens effects at first glance.

And yet even the consideration of galaxies rather than stars did not raise the chances of a gravitational lens sufficiently. At least, for over forty years after Zwicky's suggestion, no case of a gravitational lens anywhere in the sky forced itself on astronomers' attention.

This was despite the fact that a number of astronomi-

164

cal discoveries were made that seemed to increase the chances of a gravitational lens, or to make it easier to observe one.

For instance, Zwicky himself had pointed out that novas come in two types. Ordinary novas represent relatively minor gas explosions on the surface of white dwarfs. There are, however, "supernovas," which represent the explosion of most or all of a star. A supernova can shine, for a brief period, with the luminosity of billions of ordinary stars, and since a supernova rivals the brightness of an entire galaxy, it can be seen at a distance as far away as a galaxy can be seen.

By now over 400 supernovas have been detected in distant galaxies (and in 1987, there was one in the nearest of them, the Large Magellanic Cloud). Think if the light of one of them (a point source for all its brilliance) passed through an intense gravitational field on the way to us. It is conceivable we might see *two* supernovas explode close together at more or less the same time (one distorted light path might be longer than the other and therefore take a bit longer to reach us) and rise and fall with the same rhythm while displaying the same spectrum. It would be an obvious case of a gravitational lens.

However, the number of supernovas is far smaller than the number of galaxies, and each supernova is a temporary phenomenon so that the lens distortion would be equally temporary instead of being a virtually permanent feature of the sky as other gravitational lenses might be. In any case, no such distorted supernova has yet been seen.

Then, too, in 1969, pulsars were discovered. These are neutron stars that exist as a result of supernovas that have exploded in the past. They have all the mass of an ordinary star compressed into a sphere no more than the size of a small asteroid. The gravitational intensity near its surface is some millions of times as great as that near a white dwarf.

165

In addition, astronomers have become convinced that black holes exist. These represent even more monstrously condensed pieces of matter and can have, in their vicinity, gravitational fields even more intense than that of a neutron star.

Therefore a galaxy need not have its light interfered with by another galaxy. The light may pass near a neutron star or a black hole and be wrenched out of its straight-line path more sharply than might be managed by the gravitational field of a mere galaxy. The focal length would thus be shortened, and again, we would have a greater chance to see the lens effects.

The most important discovery of all, however, came in 1963, when quasars were detected. What had seemed to be ordinary and undistinguished stars of our own Galaxy raised suspicion by being the sources of detectable amounts of radio waves. Closer examination showed they had enormous red-shifts and had to be ultradistant.

Nowadays some 2,000 quasars are known. They are actually galaxies with extraordinarily bright and active centers. Because they are so far away, we see only those centers, which have the appearance of stars, and we do not usually see the faint haze of the rest of the galaxy.

Even the nearest quasar is a billion light-years away, farther off than any ordinary galaxy we can see. Some quasars, recently discovered, seem to be as much as 17 billion light-years away.

Clearly, then, if we're looking for gravitational lenses, we ought to concentrate on the quasars. They are so far away that the chances of having something on the line between them and us increases markedly. They are essentially point sources so the effects oughtn't to be too complicated. And there are sufficiently few quasars so that if a "double quasar" is found, the two being quite close together, it should at once become an object of suspicion. And if the spectra turn out to be more or less identical, that's it.

Actually, a photograph showing a double quasar in

the constellation Ursa Major was taken in the early 1950s. They were so close together that they seemed to overlap.

On March 29, 1979, the double quasar (known as 0957 + 561) was studied closely by scientists at Kitt Peak National Observatory, and the two quasars were found to be separated by 6 arcseconds. (By comparison, the width of the full Moon is 1865 arcseconds.) The chance of two quasars being that close together in the sky just as a result of random distribution is pretty small.

Furthermore, a study of the spectra of the two quasars showed them to be identical in all features. They have the same lines with the same relative prominence, and the same redshift, showing them to be at the same distance. The conclusion had to be that the two quasars were separate images of a single quasar, the double image being produced by a gravitational lens.

But what was serving as the lensing object?

Using their most sensitive light-detecting devices, astronomers spotted, between the double quasar and ourselves, a very distant (and therefore very dim) cluster of galaxies. Such clusters usually center about a giant elliptical galaxy that has grown at the expense of the smaller galaxies about it.

There *was* such a giant galaxy in the cluster and that giant was directly in front of the double quasar. (It did not obscure the double quasar because the quasar could be detected by radio waves that passed right through the elliptical galaxy between it and ourselves.) Undoubtedly, this elliptical galaxy was the lensing object.

Since then, seven more cases of objects that seem to be the result of gravitational lenses have been detected, though none seem to be as clear-cut as the first, and only one other has a clear lensing object (a spiral galaxy) between it and ourselves. In addition, there are ten more objects that *might* be lens effects. All seventeen objects, by the way, are quasars.

Most exciting of all, in 1987, a small object was discovered in the constellation Leo that has the shape of a tiny ring of radio radiation and shows all the earmarks of being the Einstein ring that Einstein had predicted fifty years before. It is the first one discovered.

Astronomers are naturally delighted by the beauty and rarity of the phenomenon, and by its example of a daring prediction, an ingenious search, and a jubilant finding, but it doesn't end there. There remains a great deal that may be accomplished by way of gravitational lenses.

In the first place, the mere existence of gravitational lenses supports once again the theory of general relativity which, for three quarters of a century, has met every test, and which is the only useful mathematical description we have of the Universe as a whole. Those small distorted images and that tiny ring of light assure us, once more, that we are on the right track and that we seem to be understanding the Universe.

Second, just as a glass lens can magnify an image that it brings into focus, so may a gravitation lens magnify the object it distorts. (This, too, was first suggested by Zwicky.)

This means that we have an unbelievably gigantic microscope that might show us the inner structure of a quasar, details that we might not be able to make out ordinarily. Astronomers would desperately love to have such information, for quasars seem to be a phenomenon of the youth of the Universe and anything that will increase our knowledge of that youth may help us come to conclusions as to the very beginnings of the galaxies, and of the Universe itself.

Then, too, as I said earlier, when a gravitational lens bends a beam of light on one side or another, splitting it into several different beams, those beams follow different paths, and one path may be longer than another. Since the curvature is, at best, slight, the paths are not very different in length proportionately, but this difference can be calculated from the geometry of the situation even if we don't know the actual distance of the quasar being distorted or the lens doing the distorting.

It may turn out, for instance, that one path is one billionth longer than another path. That's not much, but if the total length of the journey from the quasar to us is 5 billion light-years, then one beam will reach us five years after the other.

Since both beams have already arrived, however, how can we know which arrived first and by how much?

If the beams were steady in intensity, we would be out of luck. Quasars, however, sometimes show variations in their brightness. If one of the images of a multiple quasar suddenly brightens, we need simply wait for the others to brighten as their beams arrive. From the geometry and from the time lag, astronomers could calculate the distance of the quasar far more accurately than can be done by any other method.

From the distances of several quasars, determined in this fashion, and from the value of their redshifts, astronomers can calculate the value of "Hubble's constant"—the rate at which distance increases with the size of the redshift. This constant is now known only very roughly and there is much dispute over it. A good value will enable us to get a truer picture of the size and age of the Universe than anything we have now.

Then, too, there is the nature of the lensing material between the quasars and ourselves. Most of the few cases of gravitational lenses so far pinned down have nothing visible between them and us.

Perhaps the light passes by a neutron star or a black hole which we cannot possibly observe directly at that distance, but whose presence we might deduce from the distortion of the quasar.

More important still, astronomers wonder if there might not be a large quantity of mass in the Universe which, for one reason or another, we can't detect and don't take into account. That "missing mass" may account for the way in which galaxies rotate, or clusters of galaxies hang together. It may even indicate that the Universe is closed and will someday collapse, instead of expanding forever.

The bending of the light by quasars may give us a hint as to the nature of the missing mass, its location, and its quantity.

Gravitational lenses may also explain certain puzzles that hound astronomers today. There are cases of quasars with high redshifts being in apparently close association with objects of much lower redshifts. There are cases of radio sources that seem to be separating at speeds faster than light. It might be possible to find explanations of such anomalies by making use of the gravitational lensing phenomenon.

There are also thoughts about "strings," which would be folds in the space-time continuum that formed at the start of the Universe and represent exceedingly long, nearly one-dimensional objects of enormous mass. The light of a quasar that happens to intersect such a string would have the light on either side of the string bent far more than by any other lensing agent. The focal length would decrease and the two images of the quasar would seem to be separated in space to a far greater extent than anything we have so far seen.

As a matter of fact, two quasars were found separated by 157 arcseconds. They had similar spectra, and for a while, astronomers thought they might have the first piece of evidence in favor of the existence of strings. However a closer look at the spectra showed that they were not similar enough. The two quasars were, quite clearly, two *different* quasars.

In order to get the full benefit of gravitational lenses, to be sure, astronomers must find as many as possible and so some of them are eagerly planning sweeps of the sky in a general search.

12
The Secret of the Universe

Paradoxes, in the sense of self-contradictory statements, have always irritated me. It's my firm belief that the Universe works in such a way that self-contradictions do not occur. If, then, it seems we have a paradox, it can only be because we have perversely insisted on saying something we should not say.

Here, for instance, is an example of a paradox. Suppose that a certain town contains only one barber and that he shaves every man in the town except those who shave themselves. The question is: Who shaves the barber?

The barber cannot shave himself because he shaves only those who do not shave themselves. On the other hand, if he doesn't shave himself, he is bound, by the terms of the statement, to shave himself.

The paradox only arises, however, if we insist on making statements that already possess the seeds of self-contradiction. The proper way of making a *sensible* statement about the situation is to say: "The barber shaves himself and, in addition, every other man in the town except those who shave themselves." Then there is no paradox.

Here's another one: A certain despotic monarch has decreed that anyone crossing a certain bridge must declare his destination and his purpose for going there. If he lies, he must be hanged. If he tells the truth, he must be left in peace.

A certain man, crossing the bridge, is asked his destination and answers, "I am going to the gallows for the purpose of being hanged."

Well, then, if he is now hanged, he was telling the truth and should have been left in peace. But if he is left in peace, he told a lie and should have been hanged. Back and forth. Back and forth.

Again, that answer must be anticipated and must be ruled out of bounds or the decree is senseless. (In real life, I imagine the despotic monarch would say, "Hang him for being a wise guy," or "He hasn't told the truth till after he was hanged, whereupon you can allow his corpse to go in peace.")

In mathematics, there *is* the tendency to forbid the sources of paradox. For instance, if division by zero were allowed, it could be easily proved that all numbers of any kind were equal. So, to prevent that, mathematicians arbitrarily forbid division by zero, and that's all there is to that.

More subtle paradoxes in mathematics have their uses since they stimulate thought and encourage the increase of mathematical rigor. Back in 450 b.c., for instance, a Greek philosopher, Zeno of Elea, advanced four paradoxes which all tended to show that motion, as it was sensed, was impossible.

The best known of these paradoxes is usually referred to as "Achilles and the Tortoise," and this is the way it goes:

Suppose that Achilles (the fleetest of the Greek heroes involved in the siege of Troy) can run ten times as fast

172

as a tortoise, and suppose the two take part in a race, with the tortoise given a ten-yard head start.

In this case, it can be argued that Achilles cannot possibly overtake the tortoise, for by the time Achilles has covered the ten yards between himself and the tortoise's original position, the tortoise has advanced one yard. When Achilles covers the additional yard, the tortoise has advanced another tenth of a yard, and by the time Achilles runs that distance, the tortoise had advanced a hundredth of a yard, and so on forever. Achilles comes ever closer, but he can't ever quite catch up.

The reasoning is impeccable, but we all know that, in actual truth, Achilles would quickly pass the tortoise. In fact, if two people, A and B, are having a race, and if A can run faster than B by even the smallest measure, A will eventually overtake and pass B, even if B has a very large (but finite) head start, provided both parties travel at a constant best speed for an indefinitely prolonged period.

There's the paradox, then. Logical reasoning shows that Achilles cannot overtake the tortoise, and simple observation shows that he can and does.

This stumped mathematicians for two thousand years, partly because it seemed to be taken for granted that if you have an infinite series of numbers, such as $10 + 1 + \frac{1}{10} + \frac{1}{100} \ldots$, then the sum must be infinite and the time it takes to cross a distance represented by such numbers must also be infinite.

Eventually, though, mathematicians realized that this apparently obvious assumption—that an infinite set of numbers, however small, must have an infinite sum—was simply wrong. The Scottish mathematician James Gregory (1638–1675) is usually given credit for making this clear by around 1670.

In hindsight, this is surprisingly easy to show. Consider the series $10 + 1 + \frac{1}{10} + \frac{1}{100} \ldots$ Add 10 and 1 and you have 11; add $\frac{1}{10}$ to that and you have 11.1; add $\frac{1}{100}$ to that and you have 11.11; add $\frac{1}{1000}$ to that and you

173

have 11.111. If you add an infinite number of such terms, you end up with 11.111111 ... But such an infinite decimal number is only 11⅑ in fractions.

Consequently, the entire infinite set of ever-decreasing numbers that represents the lead of the tortoise over Achilles has a total sum of 11⅑ yards and Achilles overtakes the tortoise in the time it takes him to run 11⅑ yards.

An infinite series with a finite sum is a "converging series," and the simplest example, by my judgment, is 1 + ½ + ¼ + ⅛..., where each term is half the one before. If you start adding together the terms of such a series, you will have no trouble convincing yourself that the sum of the entire infinite set is, simply, 2.

An infinite series with an infinite sum is a "diverging series." Thus, the series 1 + 2 + 4 + 8 ... clearly grows larger and larger without limit, so that the sum can be said to be infinite.

It isn't always easy to tell if a series is diverging or converging. For instance, the series 1 + ½ + ⅓ + ¼ + ⅕ ... is diverging. If you add the terms, the sum grows continually larger. To be sure, the growth in the value of the sum becomes slower and slower, but if you take enough terms, you can get a sum that is higher than 2, or 3, or 4, or any higher number you care to name.

I believe that this series is the most gently diverging it is possible to have.

I learned about converging series, if I recall correctly, in my high school course in intermediate algebra, when I was fourteen, and it really struck me in a heap.

Unfortunately, I am not a natural mathematician. There were those who, even in their teenage years, could grasp truly subtle mathematical relationships—the work of men like Galois, Clairaut, Pascal, Gauss, and so on—but I was not one of them by several light-years.

I struggled with converging series and managed to see

174

something in a vague, unsystematic way and now, over half a century later, with a great deal more experience, I can present those teenage thoughts in a much more sensible way.

Let's consider the series $1 + \frac{1}{2} + \frac{1}{4} + \frac{1}{8} + \frac{1}{16} \ldots$ and try to find a way of representing them by something we can easily visualize. Imagine a series of squares, for instance, the first one being 1 centimeter on each side, the second $\frac{1}{2}$ centimeter, the third $\frac{1}{4}$ centimeter, the fourth $\frac{1}{8}$ centimeter, and so on.

Imagine them all shoved tightly together so that you have the largest square on the left, the second largest adjoining it on the right, then the third largest, the fourth, and so on. You have a line of an infinite number of smaller and smaller squares, side by side.

All of them taken together, *all* of them, would stretch across a total length of 2 centimeters. The first one would take up half the total length, the next one would take up half of what's left, the next half of what now remains, and so on *forever*.

Naturally, the squares become extremely tiny very rapidly. The twenty-seventh square is roughly the size of an atom, and once it is placed, all that is left of the 2-centimeter total is about the width of another atom. Into the second atom's-width, however, an infinite number of further squares still rapidly decreasing in size is squeezed.

The twenty-seventh square is roughly $\frac{1}{100,000,000}$ of a centimeter on each side, so let's imagine that it and all the squares that follow are magnified a hundred million times. The twenty-seventh square would now appear to be 1 centimeter on each side, followed by another square that was $\frac{1}{2}$ centimeter on each side, followed by one that was $\frac{1}{4}$ centimeter, and so on.

In short, the magnification would produce a series just equal, both in size and in number of squares, to the one we began with.

What's more, the fifty-first square is so small that it is

only about the width of a proton. Nevertheless, if *it* were expanded to a 1-centimeter square, it would have a tail of still smaller squares equal both in size and in number to what we had at the beginning.

We could keep this up *forever*, and never run out. No matter how far we went, millions of ever-decreasing squares, trillions, duodecillions, we would have left a tail precisely similar to the original. Such a situation is said to show "self-similarity."

And all of it, *all* of it, fits into a 2-centimeter width. Nor is there anything magic about the 2-centimeter width. It can all be made to fit into a 1-centimeter width, or a ¹⁄₁₀-centimeter width—or the width of a proton, for that matter.

It's no use trying to "understand" this in the same sense that we understand that there are 36 inches in a yard. We have no direct experience with infinite quantities and we can't have. We can only try to imagine the consequences of the existence of infinite quantities, and the consequences are so utterly different from anything we *can* experience that they "make no sense."

For instance, the number of points in a line is a higher infinity than that of the infinite number of integers. There is no conceivable way in which you can match those points with numbers. If you were to try to arrange the points in such a way as to line them up with numbers, you would invariably find that some points had no numbers attached to them. In fact, an infinite number of points would have no numbers attached to them.

On the other hand, you can match the points in a line 1 centimeter long with the points in another line 2 centimeters long so that you must conclude that the shorter line has as many points as the longer one. In fact, a line 1 centimeter long has as many points as can be squeezed into the entire three-dimensional Universe. You want that explained? Not by me, and not by anyone. It can be proven, but it can't be made to "make sense" in the ordinary way.

Let's get back to self-similarity. We can find it not only in a series of numbers, but in geometrical shapes. In 1906, for instance, a Swedish mathematician, Helge von Koch (1870–1924), invented a kind of supersnowflake. This is how he got it.

You start with an equilateral triangle (all sides equal), divide each side into thirds and construct a new smaller equilateral triangle on the middle third of each of the sides. This gives you a six-pointed star. You then divide each of the sides of the six equilateral triangles of the star into thirds and build a new, still smaller equilateral triangle on each middle third. You now have a figure rimmed by eighteen equilateral triangles. You divide each of the sides of these eighteen triangles into thirds— and so on and so on, *forever*.

Naturally, no matter how large your original triangle and how meticulous your draughtsmanship, the new triangles quickly become too small to draw. You have to draw them in imagination and try to work out the consequences.

If, for instance, you built up the supersnowflake forever, the length of the perimeter bounding the snowflake at each stage forms a diverging series. In the end, therefore, the perimeter of the snowflake is of infinite length.

On the other hand, the area of the snowflake at each stage forms a converging series with a finite sum. That means even at the end, with an infinite perimeter, the snowflake has an area of no more than 1.6 times the original equilateral triangle.

Suppose now that you study one of the relatively large triangles on one of the sides of the original triangle. It is infinitely complex as smaller and smaller and smaller triangles sprout off it, without end. If you take one of those smaller triangles, however, one so small that it can be seen only under a microscope, and imagine it expanded for easy viewing, it is just as complex as the larger triangle. If you consider a still smaller and an even smaller one, indefinitely, the complexity does not decrease. The supersnowflake shows self-similarity.

Here is another example. Imagine a tree whose trunk is divided into three branches. Each of those branches divides into three smaller branches and each of those smaller branches divides into three still smaller branches. You can easily imagine a real tree having a branching arrangement like that.

However, to have a mathematical supertree, you must imagine that all the branches keep dividing into three smaller branches *forever*. Such a supertree also shows self-similarity, and each branch, however small, is as complex as the entire tree.

Such curves and geometrical figures were called "pathological" at first because they didn't follow the simple rules that govern the polygons, circles, spheres, and cylinders of ordinary geometry.

In 1977, however, a French-American mathematician, Benoit Mandelbrot, began to study such pathological curves systematically and showed that they didn't fit even the most fundamental properties of geometric figures.

We are all taught, as soon as we are exposed to geometry, that a point is zero-dimensional, a line is one-dimensional, a plane is two-dimensional, and a solid is three-dimensional. Eventually, we may learn that if a solid possesses duration and exists in time, it is four-dimensional. We may even be taught that geometers can handle still higher dimensions as a matter of course.

All these dimensions, however, are whole numbers—0, 1, 2, 3, and so on. How can we have anything else?

Mandelbrot, however, showed that the boundary of the supersnowflake was so fuzzy and made such sharp turns at every point that there was no use considering it to be a line in the ordinary sense. It was something that was not quite a line and yet not quite a plane either. It had a dimension that was *in between* 1 and 2. In fact, he showed that it made sense to consider its dimension to be equal to the logarithm of 4 divided by the logarithm of 3. This comes to about 1.26186. Thus, the boundary of the supersnowflake has a dimension of just over 1¼.

Other such figures also had fractional dimensions, and because of this, they came to be called "fractals."

It turned out that fractals are not pathological examples of geometric shapes that were dreamed up through the fevered imaginations of mathematicians. Rather, they are closer to the real objects of the world than were the smooth, simple curves and planes of idealized geometry. It is these latter that are the products of imagination.

In consequence, Mandelbrot's work became more and more important.

Now let's change the subject slightly. Several years ago, I had occasion to hang about Rockefeller University now and then and I met Heinz Pagels there. He was a tall fellow with white hair and a smooth, unlined face. He was exceedingly pleasant and bright.

He was a physicist and knew much more about physics than I did. This was no surprise. Everyone knows more than I do about something or other. It also seemed to me that he was more intelligent than I was.

You might think, if you share the general opinion that I have a giant ego, that I would hate people who seemed more intelligent than I do, but I don't. I have discovered that people more intelligent than I am (and Heinz was the third of the sort I had met) are extremely kindly and pleasant, and besides I have found that if I listen carefully to them, I am stimulated sufficiently to work up useful ideas; and ideas, after all, are my stock in trade.

I remember that in our first conversation, Heinz talked about the "inflationary Universe," a new idea to the effect that the Universe, in the first instant after its formation, expanded at enormous speed, thus ex-

179

plaining some points that bothered astronomers who had assumed the initial moments of the Big Bang to be noninflationary.

What particularly interested me was Heinz's statement that, according to this theory, the Universe started as a quantum fluctuation of the vacuum, and was thus created out of nothing.

This got me excited because in the September 1966 issue of *Fantasy and Science Fiction*, years before the inflationary Universe had been thought up, I had published an essay called "I'm Looking Over a Four-leaf Clover," in which I suggested that the Universe was created out of nothing at the time of the Big Bang. In fact, a key statement in the essay was my definition of what I called "Asimov's Cosmogonic Principle"—i.e., "In the Beginning, there was Nothing."

This does *not* mean I anticipated the inflationary Universe. I just get these intuitional thrusts, but I lack the ability to carry through. Thus, at fourteen, I had the vague intuitional notion of self-similarity in connection with converging series, but neither then, nor at any time thereafter, could I possibly have duplicated what Mandelbrot did. And though I grasped the concept of creation out of nothing, I couldn't in a million years have worked out the detailed theory of an inflationary Universe. (However, I'm not a total failure. I realized early on that my intuitional grasp made it possible for me to write science fiction.)

I met Heinz periodically thereafter, all the more so after he became the director of the New York Academy of Sciences.

One time, a bunch of us, including Heinz and myself, were sitting about discussing this and that, and Heinz raised an interesting question.

He said, "Is it possible, do you suppose, that someday, all the questions of science will be answered and there will be nothing left to do? Or is it impossible to get all the answers? And is there any way we can conceivably

180

decide, right now, which of these two situations is correct?"

I was the first to speak. I said, "I believe we can decide right now, Heinz, and easily."

Heinz turned to me and said, "How, Isaac?"

And I said, "It's my belief that the Universe possesses, in its essence, fractal properties of a very complex sort and that the pursuit of science shares these properties. It follows that any part of the Universe that remains understood, and any part of scientific investigation that remains unresolved, however small that might be in comparison to what is understood and resolved, contains within it all the complexity of the original. Therefore, we'll never finish. No matter how far we go, the road ahead will be as long as it was at the start, and that's the secret of the Universe."

I reported all this to my dear wife, Janet, who looked at me thoughtfully and said, "You'd better write up that idea."

"Why?" I said. "It's just an idea."

She said, "Heinz might use it."

"I hope he does," I said. "I don't know enough physics to do anything with it, and he does."

"But he might forget he heard it from you."

"So what? Ideas are cheap. It's only what you do with them that counts."

Then came July 22, 1988, when Janet and I headed out to the Rensselaerville Institute in upstate New York to conduct our sixteenth annual seminar, which on that occasion was to be centered about biogenetics and its possible side-effects—scientific, economic, and political.

Something extra was added, however. Mark Chartrand (whom I had met years ago when he was the director of the Hayden Planetarium in New York) is a perennial faculty member at these seminars, and he had brought with him a thirty-minute videocassette featuring fractals.

In just the last few years, you see, computers have become powerful enough to produce a fractal figure and slowly expand it millions and millions of times. They can do this with very complex fractals, not merely things as simple (and, therefore, uninteresting) as supersnow-flakes and supertrees. What's more, the videotape was made more brilliant with false color.

We began to watch the cassette at 1:30 P.M. on Monday, July 25, 1988.

We started with a dark cardioid (heart-shaped) figure, which had small subsidiary figures about it, and little by little it grew larger on the screen. One subsidiary figure would slowly be centered and grow larger until it filled the screen and it could be seen that it was surrounded by subsidiary figures, too.

The effect was that of slowly sinking into complexity that never stopped being complex. Little objects that looked like tiny dots grew larger and revealed complexity while new little objects formed. It *never stopped*. For half an hour, we watched it as different parts of the figure were expanded into new visions of unceasing beauty.

It was absolutely hypnotic. I watched and watched, and after a while, I simply couldn't withdraw my attention. The whole was as close as I ever came, or could come, to *experiencing* infinity, instead of merely imagining it and talking about it.

When it was over, it was a wrench to come back to the real world.

Afterward, I said dreamily to Janet, "I'm sure I was right in what I said to Heinz that time. That's the Universe and science—endless—endless—endless. The job of science will never be done, it will just sink deeper and deeper into never-ending complexity."

Janet frowned. "You still haven't written up that idea, though, have you?"

182

And I said, "No, I haven't."

But while we were at the Institute, we were isolated from the world. There were no newspapers, no radio, no television, and we were too busy with the details of the seminar to worry about that.

It wasn't until we got back to our apartment on the twenty-seventh, and I was leafing my way through the accumulated newspapers that I found out what had happened.

While we were in Rensselaerville, Heinz Pagels was attending an extended meeting on physics in Colorado. Pagels was also an enthusiastic mountain climber, and during the weekend break, on Sunday, July 24, he climbed Pyramid Peak, 14,000 feet high, along with a companion. He had lunch there and at 1:30 P.M. (just twenty-four hours before I started watching the videocassette), he started down the mountain.

He stepped on a rock that was loose. It trembled under him and he lost his balance. He went sliding down the mountainside and was killed. He was forty-nine years old.

I, totally unprepared, turned to one of the obituary pages and saw the scare headline. It was a bad and unexpected shock and I'm afraid I must have cried out in unhappiness for Janet came running and read the obituary over my shoulder.

I looked up at her, sorrowfully, and said, "Now he'll never have a chance to use my idea."

So now, at last, I have written it up. Partly, this was so I could say something about Heinz, whom I so admired. And partly, it's because I wanted to put the notion on paper so that (just possibly) someone—if not Heinz, *some-one*—might be able to use it and do something with it.

After all, I can't. Just getting the idea represents my total ability. I can't move an inch beyond that.

Part III
Here at Home

13
The Salt-Producers

When I was young, in the far-off distant days before antibiotics, we had a home remedy for any cut, scrape, or abrasion. In order to prevent infection, we smeared the spot with tincture of iodine—that is, a solution of iodine in alcohol.

It was ground into me that iodine was the universal anti-infection treatment. It stung on application (ouch, ouch) but that was good for I always felt, as a child, that the stinging was a sign that all the germs were being killed.

But time has passed and home remedies have changed. My dear wife, Janet, is, of course, an M.D., and so she is up on all the latest anti-infection stuff. Her greatest happiness in life is doctoring me for any minor problem I may have. (It's not *my* greatest happiness, but I love her dearly and am willing to endure the inconvenience if it will make her happy.) In any case, she plasters me with a variety of ointments and lotions and antibiotic creams.

However, in my medicine cabinet I insist on having a little bottle of tincture of iodine, and anytime I can hide a cut or scrape or abrasion from dear Janet's eagle eye, I smear it with iodine so that I can feel that healthy germ-killing sting.

And because I used it the other day for exactly that purpose (and was caught by Janet, who gave me Lecture 3-A on the subject), I thought that, since I occasionally write an essay on one or another of the chemical elements, I ought to do one on iodine. Here goes—

We won't begin with iodine, however, but with hydrochloric acid.

The medieval alchemists discovered three strong mineral acids which proved far more powerful in bringing about chemical changes than was the comparatively weak acetic acid. Acetic acid is the active principle of vinegar and was the best the ancients could do in this direction. Two of the strong mineral acids to be discovered were sulfuric acid and nitric acid, and the third was hydrochloric acid.

We don't really know for sure who discovered the acid or when, but the first clear description of the preparation of hydrochloric acid came in a book written about 1612 by the German alchemist Andreas Libau (1560–1616), better known by the Latinized version of his name, Libavius.

Naturally, he didn't call it hydrochloric acid, since that name appeared only in the nineteenth century, after modern chemical nomenclature had been invented. He called it "spiritus salis," which is Latin for "spirit of salt." He called it that because he formed it from salt and water, heated in the presence of clay. It was a "spirit" because it was easily vaporized, whereas salt itself was not.

In later years, hydrochloric acid was called "marine acid" because the salt from which it was formed could be obtained from seawater. The salt could, as well, be obtained from brine (salt-saturated water found here and there on land) so that it was also called "muriatic acid" from the Latin word for "brine."

We now skip a century and a half—

In 1774, the Swedish chemist Karl Wilhelm Scheele (1742–1786) heated muriatic acid with a mineral called "pyrolusite," which contained what we would today call manganese dioxide.

When he did that, he produced a vapor with a suffocating odor that made him cough and which he found "oppressive." He worked with it anyway (and with various other poisonous vapors, which may account for his early death at the age of forty-four), and found that the vapor dissolved slightly in water, attacked metals, and bleached the green out of leaves and the color out of flowers.

Here is what had happened. Muriatic acid has molecules made up of one hydrogen atom and one chlorine atom (which is why it is now called hydrochloric acid). Oxygen from the manganese dioxide combined with the hydrogen atom, liberating the chlorine atom. The chlorine atoms so liberated combined in pairs to form chlorine molecules and these formed vapors of the gaseous element chlorine. In short, Scheele had discovered chlorine.

Scheele, however, although one of the great chemists of history was, by all odds, the unluckiest. He was the first to isolate about half a dozen chemical elements, but got credit for not a single one. Either his publication was delayed, allowing someone else to get in ahead of him, or the credit went to a coworker, or *something*.

In the case of chlorine, his report was first and there was no coworker to confuse things, but Scheele lost out anyway because he did not recognize the vapors he had produced as being an element. He accepted a theory, evolved in 1700, which considered combustion to involve the loss of something called "phlogiston." He therefore called his vapor, produced by the combustion of muriatic acid, "dephlogisticated muriatic acid."

Just about the time that Scheele was isolating chlorine, the British chemist Joseph Priestley (1733–1804) discovered oxygen. (Scheele had discovered it first, but his paper on the subject was disgracefully delayed by the publisher and he lost the credit.) The French chemist Antoine Laurent Lavoisier (1743–1794) had, by 1778, showed that combustion is due not to the loss of phlogiston, but to the gain of oxygen.

This meant that Scheele had not deprived muriatic acid of phlogiston but had added oxygen to it. Dephlogisticated muriatic acid came to be called "oxymuriatic acid." Scheele's vapor was still thought not to be an element but to be a loose combination of muriatic acid and oxygen.

Lavoisier was of the opinion that all acids contain oxygen (which was why he called the gas "oxygen," from the Greek words for "acid producer"). He therefore considered both muriatic acid and oxymuriatic acid to contain oxygen.

The matter was investigated by the British chemist Humphry Davy (1778–1829), and by 1808, he had come to the conclusion that muriatic acid did *not* contain oxygen; that it was the hydrogen content that combined with oxygen; and that Scheele's vapor did not contain oxygen either but was an element. That is, the vapor could not be broken down to anything simpler and more fundamental.

Because the vapor was greenish in color, Davy called it "chlorine," from the Greek word for "green." He maintained that muriatic acid was composed of molecules that contained hydrogen and chlorine, so that it came to be called "hydrogen chloride" or "hydrochloric acid." Because all this was correct, it is Davy who is given credit for the discovery of chlorine, and not Scheele.

Not everyone was convinced by Davy's reasoning, however. In particular, the Swedish chemist Jöns Jakob Berzelius (1779–1848) continued to maintain that Davy's chlorine was actually an oxide of some sort. Berzelius was the most prominent chemist of his day and was virtually the dictator of chemical belief. As long as he refused to accept Davy's thesis, that thesis remained uncertain.

That brings us to a French chemist, Bernard Courtois (1777–1838).

Courtois was the son of a manufacturer of saltpeter in Dijon, and while he received a good chemical education, he took over his father's business in 1804 and was what we might call an "industrial chemist."

Manufacturing saltpeter was not a trivial occupation at this time. Saltpeter, or "potassium nitrate" in chemical terminology, is an essential component of gunpowder. The other components, sulfur and charcoal, were easily and abundantly available, but saltpeter was not.

Courtois took over the firm in the middle of the Napoleonic wars, and gunpowder was desperately needed by France. Great Britain was the chief enemy and the British Navy controlled the sea. The best sources of saltpeter were abroad, and as these had become unavailable, Courtois's business was in deep trouble.

He attempted to get the necessary saltpeter from seaweed obtained from the Channel shores of France. He burned the seaweed and dissolved the saltpeter out of the ash with water. The water, however, also dissolved various components of the ash that were undesirable impurities. To get rid of these impurities, Courtois added a bit of sulfuric acid and this precipitated the impurities and Courtois could filter them off.

What Courtois didn't know, however, and in fact, what no one at the time knew, was that the seaweed contained tiny quantities of compounds containing a hitherto unknown substance.

Toward the end of 1811, Courtois seems to have added a bit too much sulfuric acid and, to his surprise, obtained a violet vapor that had an irritating odor much like chlorine and that solidified on cold objects to form nearly black crystals.

Courtois studied these crystals and found that they would combine with hydrogen, phosphorus, and metals. They also combined with ammonia to form an explosive substance (one that was not well behaved enough to substitute for gunpowder, however). It did not combine readily with carbon or oxygen. In all these respects, the new substance rather resembled chlorine.

Courtois also found that heating the crystals strongly did not break them down to anything simpler. He therefore suspected he had a new element, but he was forced to drop his investigation. For one thing, he had no money and no time. It took all his efforts to support himself and his family with his near-bankrupt factory.

For another thing, he doubted his own chemical ability to handle a problem as difficult as that of determining whether a strange new substance was an element or not. He discontinued his investigations, therefore, and did not even publish those observations he had made. In fact, in his whole life, Courtois never published a scientific paper.

In July 1812, however, he passed on the information he had gathered to two local chemists of repute—Nicholas Clément (1778–1841) and his collaborator, Charles Bernard Desormes (1777–1862). They continued the research, and when they reported their work to the Institut de France on November 19, 1813, they gave Courtois full credit for the discovery.

Clément, having studied the properties of the new substance in great detail, found, like Courtois, that they resembled those of chlorine. He therefore announced that the vapor was a new element.

By that time, other chemists were working with the new substance, including Humphry Davy and a French chemist, Joseph Louis Gay-Lussac (1778–1850). Both agreed that the substance was a new element, resembling chlorine. Gay-Lussac named the element from the violet color of its vapor. The Greek word for "violet" is "ion" and "violetlike" is "iodes." Adopting the "-ine" ending of chlorine, we end with "iodine."

There was some argument over priority between Davy and Gay-Lussac, mostly as a matter of politics. After all, Davy was the preeminent British chemist and Gay-Lussac the leading French chemist, and the two nations were in a deadly war. However, the discovery went to neither but is universally credited to Courtois, which is as it should be.

The data which showed iodine to be an element was so strong that its elementary nature was soon universally accepted by chemists. The fact that it was an element made it seem far more likely that the very similar substance, chlorine, was an element also. By 1820, even that very stubborn and dictatorial chemist, Berzelius, gave in.

None of this did Courtois any financial good, however. His saltpeter business finally failed and he tried to make a living out of preparing and selling iodine but that brought in very little. In 1831, he was awarded a prize of 6,000 francs for his discovery, but that was soon spent, and in 1838, he died in total poverty.

The story is not yet over, for we must consider yet another French chemist, Antoine Jérôme Balard (1802–1876), who, in 1824, was a young assistant at a pharmaceutical school in Montpellier.

He was interested in trying to find new sources of iodine and, for the purpose, was studying different types of plants that grew in salt marshes. He extracted the ash of these plants and boiled them down, and then discovered that when he added certain chemicals, the liquid turned brown.

He investigated the brown material and finally obtained a reddish liquid that, like chlorine and iodine, had a choking, cough-producing odor. He studied its properties and found them to be more or less midway between those of chlorine and iodine. The liquid was darker than chlorine in color, but lighter than iodine. It had the same chemical properties, but seemed more chemically active than iodine and less chemically active than chlorine.

For a while, Balard thought that what he had was a compound of chlorine and iodine—"iodine chloride," so to speak—but he was unable to break down the new substance and to produce either iodine or chlorine from it. He therefore announced it as a new element, and named it "muride" from the brine in which the plants producing it had grown.

The matter was studied by a group from the French Academy, including Gay-Lussac, and they refused to accept the name, perhaps because it reminded them of the outmoded term "muriatic acid." Instead, they turned to the strong and unpleasant odor of the vapor and called it "bromine" from the Greek word for "stench."

The discovery of bromine had the usual number of near-misses that most discoveries have.

In 1825, a young German chemistry student, Carl Löwig (1803–1890), obtained a red liquid from the dissolved contents of a salt spring. He brought it with him into the laboratories of a notable German chemist, Leopold Gmelin (1788–1853), and Gmelin at once suggested that he study the liquid carefully to see if it was a new element.

Löwig did so, but before he had come to a decision, Balard's paper announcing the discovery of bromine had appeared. However, Löwig went on to a long and successful chemical career, much of it engaged in the study of bromine compounds, so that his near-miss was not particularly tragic.

More dramatic was the fate of still another young German chemist, Justus von Liebig (1803–1873). He had obtained a sample of a red liquid even before Löwig had and he had carefully tested it, put it in a vial, labeled it, and added it to the reagent shelf in his laboratory.

When the news of the discovery of bromine came out, Liebig rushed to his shelf, got down the vial, tested its contents again, and found that he had had pure bromine sitting there for many months. Then he stared at the label he had written—which said "iodine chloride." Liebig, however, went on to become one of the great organic chemists of all time so here, too, the tragedy is not very great.

It was obvious that the properties of bromine lie between that of chlorine and iodine. A German chemist, Johann Wolfgang Döbereiner (1780–1849), reduced this to numbers, however. He noticed that the atomic weight of bromine is just about midway between that of chlorine and iodine. If the atomic weights of chlorine and iodine are added and the sum is divided by 2, the answer is 81. The atomic weight of bromine is about 80.

Döbereiner noticed the same thing in connection with other groups of three elements with similar properties— for example, calcium, strontium, and barium. When the atomic weights of calcium and barium are added and the sum is divided by 2, the answer is about 89. The atomic weight of strontium is about 88.

Then there are sulfur, selenium, and tellurium, which have similar properties. The atomic weights of sulfur and tellurium, when added together and divided by 2, have an average of about 80, and the atomic weight of selenium is about 79.

Döbereiner suggested, in 1829, that the elements exist in groups of three, which he called triads.

This was important. By Döbereiner's time, over fifty elements were known and their properties varied all over the lot. They were a disorderly bunch and it is virtually an instinct with scientists to try to find order wherever they can.

Döbereiner's triads were the first suggestion of order in connection with the elements, and that suggestion had been advanced, not surprisingly, soon after bromine had been discovered, since chlorine, bromine, and iodine made up the most noticeable and clear-cut example of a triad.

The notion of triads did not flourish, however, for several reasons. First, though there were over fifty elements known, thirty remained yet to be discovered and their absence left holes that made any orderly arrangement of elements difficult. Second, the atomic weights of many elements were not yet correctly known so that the notion of triads would fail even where they existed and might show up where they did not exist. Third, the true order of the elements is so subtle that something as simple as triads will not do.

Nevertheless, Döbereiner's suggestion stirred the imagination of chemists and others began to tackle the job of finding order in the apparently disorderly mob of elements.

In 1869, forty years after the idea of the triads was suggested, the Russian chemist Dmitri Ivanovich Mendeléev (1834–1907) finally worked out the true order of the elements in the form of a "periodic table." This was the greatest chemical discovery between Lavoisier's elucidation of the nature of combustion in 1778, and the discovery of radioactivity by the French chemist Antoine Henri Becquerel (1852–1908) in 1896.

The triads were sufficient to make it possible to refer to some groups of elements by a general name. Thus, chlorine combines with sodium to form sodium chloride, or ordinary table salt. Bromine and iodine combine with sodium to form sodium bromide and sodium iodide, respectively, and each of these compounds resembles salt in many ways. For this reason, chlorine, bromine, and iodine are referred to as the "halogens," from Greek words meaning "salt-producers."

Are chlorine, bromine, and iodine the only halogens there are?

The answer is no! Chlorine has the smallest atomic weight of the three, but there is a halogen with a still smaller atomic weight that is called "fluorine."

Fluorine carries on the progression. Whereas iodine is a solid at ordinary temperatures, with a boiling point at 184° C; bromine is a liquid, with a boiling point at 59° C. Chlorine and fluorine are both gases at ordinary temperatures with a boiling point of $-35°$ C for chlorine and $-187°$ C for the smaller fluorine.

Bromine is more chemically active than iodine, while chlorine is more active still, and fluorine is most active of all. Fluorine is, in fact, the most active of all the chemical elements—combining most eagerly with other atoms, and being pried loose with the greatest difficulty.

The existence of fluorine in certain compounds seemed certain rather early on, but it was a long time before anyone actually managed to pry it loose and prepare it as an elemental gas. The task was carried through by the French chemist Ferdinand Frédéric Henri Moissan (1852–1907), in 1886, and in 1906 he received a Nobel Prize in chemistry for the feat.

There is a fifth halogen also, at the other end of the list, beyond iodine. This came to be known in 1914, when the British physicist Henry Gwyn-Jeffreys Moseley (1887–1915) worked out the concept of the "atomic number."

This put the finishing touch on the periodic table by giving each element an integral number. This meant that it was known exactly where each element fit in the table. The numbers made it possible to know that between two particular known elements, there could be no other—or that there could be one, or two, new elements.

The atomic number of fluorine is 9, that of chlorine

is 17, that of bromine is 35, and that of iodine is 53. The periodic table is so arranged that there must be a fifth halogen at atomic number 85, but no element was known there.

By the time of Moseley's discovery, this was not surprising. Radioactivity had been discovered and it was clear that all the elements with atomic numbers above 83 are radioactive and have no stable isotopes.

Elements 90 and 92 (thorium and uranium, respectively) are nearly stable and therefore exist in the Earth's crust in considerable quantities. As they break down, they form other elements with atomic numbers higher than 83. These exist only in traces, but they can be detected by their radioactive radiations.

In 1931, an American chemist, Fred Allison, making use of a technique he had devised called "magneto-optics," reported having detected the fifth halogen and named it "alabamine." He was proved to be mistaken, however, and its place in the periodic table remained unoccupied. The fifth halogen continued to be referred to simply as "element 85" or as "ekaiodine." ("Eka" is Sanskrit for "one," so that eka-iodine is "one" beyond iodine.)

If element 85 cannot be detected among the breakdown products of uranium or thorium, it might be possible to make it in the laboratory. Suppose you start with bismuth, which has an atomic number of 83, and bombard it with alpha particles, which are helium nuclei and have an atomic number of 2. If some alpha particles fuse with bismuth nuclei, the combined nuclei would have an atomic number of 85.

In 1940, a team of physicists at the University of California, including the Italian-American Emilio Segrè (1905–1989), tried the experiment and thought they might have formed eka-iodine. However, World War II was raging in Europe and the chances of American involvement were increasing steadily. That particular project, then, was put to one side.

In 1947, the project was begun again and eka-iodine was formed in tiny traces. It was named "astatine" from a Greek word meaning "unstable" and unstable it was indeed. Its most nearly stable isotope was astatine-210 and it had a half-life of only 8.3 hours.

Suppose we consider only the four stable halogens: fluorine, chlorine, bromine, and iodine. In general, though with some exceptions, small atoms are more common in the Universe than large atoms are. It is not surprising, then, that in the Earth's crust, fluorine is the most common of the halogens, then chlorine, then bromine, and finally iodine.

In parts per million by weight, the figures are fluorine, 700; chlorine, 200; bromine, 3, and iodine, 0.3. The most common compounds of the halogens are those with sodium and potassium and they are quite soluble. On the whole, then, these compounds tend to be washed out of the land and into the sea. (That's why the ocean is salty.)

Iodine is, of course, far less common in the sea than chlorine is. The amount of sodium iodide in the sea is only 1/8,000 of the amount of sodium chloride. You'd have to poke through 500 tons of seawater to get out an ounce of iodine and this is not really a practical process.

This could seem frightening since iodine is essential to life. (So is chlorine, but there is plenty of that. Fluorine is useful, in tiny quantities, in tooth formation only, and bromine is, as far as we know, not essential to life.)

To be sure, iodine is needed only in small quantities. The human body contains only about 14 milligrams of iodine (about 1/2,000 of an ounce), mostly in the thyroid gland, but how do we manage to get even that small amount?

Fortunately, iodine is also essential for sea life, and the plant world of the ocean, including seaweed, filters it out of the water and concentrates it in its own tissues. (That's why Courtois managed to get iodine out of seaweed; he couldn't have done it out of seawater.)

This means that seafood contains ample iodine for our needs.

But what about inland places where seafood is not common? Again, fortunately, the salt spray of the ocean dries in the sun to tiny particles of salts, including sodium iodide, which can be blown far inland by the winds. Small amounts of iodine compounds are deposited in the soil and, therefore, are absorbed by land plants, from which it reaches us.

There are places, however, where the iodine content in the soil is consistently low—places such as the Alps, the Rocky Mountains, and the American Midwest. Iodine shortages are endemic there. The result, then, is that the thyroid grows and produces goiters with all sorts of unpleasant symptoms.

Once iodine was discovered, however, it was soon found to be helpful in the treatment of goiter. Eventually, small quantities of sodium iodide were added to the water supplies of cities that suffered from shortages. Sodium iodide could also be added in small quantities to salt, forming "iodized salt."

And iodine could also be used as an antiseptic, even today, by backward people such as myself who trust the iodine sting more than the antibiotic lotion.

14
The True Rulers

I have always found the history of ancient Greece to be a gold mine of good stories, and for some reason, I remember them all.

Consider Themistocles, for instance. He was the Athenian leader who persuaded the city to invest in a fleet while they awaited the attack of the Persians. In 480 B.C., the Persians came, swept down from the north, took Athens, and burned it. The Athenian population had fled to the islands, protected by the Athenian fleet, and now that fleet (plus ships from other Greek cities) was waiting for the Persian fleet in the narrow strait between Athens and the island of Salamis.

The titular leader of the fleet was Eurybiades of Sparta (which was then the leading military power among the Greek cities). The Spartans were faultlessly brave on land, but a little nervous at sea. Eurybiades wanted to retreat to protect Sparta, seeing that Athens had already been destroyed, but Themistocles wanted to stay and fight.

Themistocles argued his case with such pertinacity that Eurybiades, exasperated at the eloquent flow of words, raised his staff of office threateningly. Themistocles threw his arms wide.

"Strike," he said, "but *listen*."

Eurybiades decided to stay. Themistocles, in order to make sure he did not change his mind, sent a messenger to the Persian king, Xerxes, suggesting that he station ships at both ends of the strait of Salamis so as to trap the Greek fleet.

Came the morning. The Greek ships found themselves penned in, had no choice but to fight, and destroyed the Persians. The battle of Salamis was the decisive engagement of the Persian war.

After the battle, the captains of the Greek ships got together to vote on who should get the prize for achievement in their great victory. Every single captain voted for himself in first place; and every single captain voted for Themistocles in second place.

There is a story that Themistocles, now at the height of his fame, was sneered at by a Greek from some small backwoods Greek town. He said, "You would not have achieved fame had you happened to have been born in my small town." And Themistocles answered, "Nor you, had you happened to have been born in Athens."

But my favorite Themistocles story is the one in which he pointed to his infant son and said, "There is the ruler of Greece."

"That child," said someone else in amazement.

"Certainly," said Themistocles, "for Athens rules Greece, and I rule Athens, and my wife rules me, and that child rules my wife."

So let's find out who rules the Earth.

Back in the last quarter of the 1600s, a Dutchman named Anton van Leeuwenhoek (1632–1723) ground tiny lenses as a hobby. Some of his excellent creations were no larger than the head of a pin, but through them he could magnify tiny things up to 200 times, seeing them more clearly than could any other person of his time. He ground a total of 419 lenses over a fifty-year period, working right down to the end of his long life.

He was the first, in 1673, to discover one-celled organisms, too small to see without a microscope, yet as indubitably alive as the largest whale. He saw capillaries and red blood cells and yeast cells and spermatozoa.

His greatest discovery, however, came in 1683, when he observed and drew pictures of the smallest things that his best lenses could show him. He didn't know what they were, and no one else would see them for another century. However, looking at the pictures he drew, we know that van Leeuwenhoek was the first man ever to see bacteria.

Of course, that's not what van Leeuwenhoek called them. He called all the tiny living things he saw "animalcules" ("small animals" in Latin). Nowadays, we lump them together as "microorganisms" ("small animals" in Greek).

The first person who really tried to *study* the bacteria was a Danish biologist, Otto Friedrich Muller (1730–1784). His observations appeared in a book that was published posthumously in 1786.

He was the first to try to classify microorganisms generally into categories; that is, into species and genera in the fashion made popular a half-century earlier by the Swedish naturalist Carolus Linnaeus (1707–1778). However, Linnaeus worked with plants and animals easily visible to the eye and could make classifications on the basis of detailed differences and similarities of clearly seen parts.

Microorganisms, on the other hand, are tiny, and very little detail could be seen in them. Nothing much could be done except to judge them by their overall shape, particularly where bacteria were concerned. That was like trying to classify ordinary plants and animals by the

203

shadows they cast. Muller did notice, however, that some bacteria are shaped like tiny rods and some like tiny corkscrews. The former he called "bacilli" or, in the singular, "bacillus" ("small rod" in Latin); and the latter he called "spirilla" or, in the singular, "spirillum" ("small spiral" in Latin).

The term "spirilla" retains its specialized meaning, but "bacilli" is sometimes used as a synonym for bacteria generally.

It seemed unlikely, in Muller's time, that anyone would ever be able to see bacteria any more clearly than Muller did. The lenses used in microscopes refracted light to a different extent as the wavelength changed. You could bring one wavelength into focus but the others would remain out of focus and show up as obscuring rings of color about the object you were trying to see.

In 1830, however, a British lensmaker, Joseph Jackson Lister (1786–1869), succeeded in forming microscope lenses out of two different kinds of glass. Each kind reflected light differently with respect to wavelength, and if they were combined in just the right way, the color effects of one were just canceled by those of the other. It became an "achromatic lens" ("no color" in Greek).

Using "achromatic microscopes," the focus could be made sharp without obscuring rings of color and only then did it become possible to study something as small as bacteria in a meaningful way.

Then, in the 1860s, the French chemist Louis Pasteur (1822–1895) began to insist that infectious disease is the result of the spread of specific microorganisms from one person to another. This was the greatest single medical discovery of all time, and focused attention wonderfully on microorganisms.

Inspired by Pasteur's work, the German botanist Ferdinand Julius Cohn (1828–1898) became the first scientist to concentrate his life's work on bacteria. In 1872, he published a three-volume treatise on bacteria that may be considered to have founded the science of bacteriology. He went much farther than Muller did in the classification of bacteria, and was the first to describe bacterial spores and their resistance to even boiling temperatures.

He kept Muller's division of bacteria into bacilli and spirilla, but went further. He noted that some rod-shaped bacteria are longer than others. He reserved the word "bacilli" for the longer rods. For the shorter ones, he was the first to make use of "bacteria," or in the singular, "bacterium" (also "small rod" in Latin).

For some reason, "bacteria" came to be the term used most often for these microorganisms generally, although still other terms also came into use. Thus, the German pathologist Christian A. T. Billroth (1829–1894) called bacteria which had the shape of tiny spheres "cocci" or, in the singular, "coccus" ("berry" in Greek). Some varieties are "streptococcus," "staphylococcus," and "pneumococcus."

Then, too, the French biologist Charles Sedillot introduced the term "microbe" ("small life" in Greek) for all minute organisms that cause disease, putrefaction, or fermentation. "Microbe" is another word that is sometimes applied to bacteria generally.

A still more general term that came into use at the beginning of the 1800s is the one that is least applicable to bacteria specifically, but is most often used in that way by the general public. It is "germ" ("sprout" in Latin), and it can be used to indicate any tiny object from which life can spring.

Thus, the portion of a seed which contains the actual living material can be considered the germ, so that we speak of things like "wheat germ." Again, life springs from sperm and ova, so these are called "germ cells." In the developing embryo, the primitive groups of cells from which organs eventually develop are "germ layers."

As a matter of fact, Pasteur's thoughts on infectious illnesses are usually referred to as the "germ theory of disease," which is incorrect actually, since bacteria are by no means the only pathogenic organisms. Disease can also be caused by viruses, molds, protozoa, parasitic worms, and so on.

The most obvious property that separates bacteria from other cells is size. One-celled organisms that are not bacteria may be large enough to be on the edge of being visible to the unaided eye. They have to be since they must pack a great deal of functioning into their single cell. An amoeba, for instance, is about 200 micrometers ($\frac{1}{125}$ of an inch) across.

The cells that make up multicellular organisms are smaller than this. They don't have to carry a full load of material for independent life. They can share the labor with other cells. The human liver cell, for instance, is about 12 micrometers across. About 2,400 human liver cells would fit into an amoeba.

A typical bacterium may, however, be only 2 micrometers across. Bacteria are the smallest free-living bits of life there are on Earth, or possibly, the smallest there can be. The smallest bacteria we know are about 0.02 micrometers across. About 200 million of these smallest bacteria can fit into the cell of an amoeba.

(There are living objects, called viruses, that are smaller than bacteria, but none of them are free-living. They can grow and reproduce only inside a living cell.)

Where do bacteria fit in the hierarchy of life? When I was young, I learned that all life is divided into two "kingdoms," plants and animals. I gathered that bacteria were placed, a little uncomfortably, within the plant kingdom. Alternatively, plants and animals include only multicellular life, while all unicellular life make up a third kingdom called "Protista" ("first" in Greek).

To understand the present view, we must go back to 1831, when the British botanist Robert Brown (1773–1858) was the first to notice that inside ordinary cells there are tiny structures. He called these "nuclei," or in the singular, "nucleus" ("little nut" in Latin) since they are found inside the cell, like a nut inside a shell.

As it eventually turned out, the nucleus of a cell contains the genetic material that controls cell reproduction. The genetic material replicates itself during cell division and passes a more or less exact copy from parent cell to daughter cell; and in a larger sense, from parent organism to child organism.

Each complete cell in all multicellular organisms contains a nucleus, whether these organisms are plants or animals. (There are, to be sure, incomplete cells such as red blood corpuscles, which do not contain nuclei, but they are short-lived and neither grow nor divide.)

Multicellular animals and plants, therefore, can both be lumped together as being made up of nucleated cells, or "eukaryotes" ("true nucleus" in Greek). In addition, one-celled animal cells such as amoebae and one-celled plant cells such as "algae" or, in the singular, "alga" ("seaweed" in Latin) are eukaryotes.

In other words, plants and animals, including all the multicellular forms, together with the larger unicellular forms, may be considered to form a "superkingdom" of "Eukaryota."

Bacterial cells, on the other hand, do not contain nuclei. That does not mean that they do not contain genetic material. They must, for they grow and multiply. The genetic material is not sequestered in a nucleus, however, but is distributed throughout the bacterial cell. Or you might say that the bacterial cell is essentially a free-living cellular nucleus and that is why it is so small. (However, the bacterial cell also contains structures that, in eukaryotes, are contained outside the nucleus.)

Bacterial cells, and in fact any cells that do not contain a clearly delimited nucleus but have genetic material distributed through the cell, are "prokaryotes" ("before the nucleus" in Greek), and might be considered as being included in the superkingdom of "Prokaryota." In this way, we divide life into two parts: bacteria and everything else.

The word "prokaryote" implies that bacteria are more primitive than the eukaryotes and, therefore, that they may have evolved and existed before the eukaryotes did.

If we go back into the fossil record, we find that we are dealing with relics of multicellular creatures, on the order of our own complexity, and many of them are quite large. From their resemblance to creatures living today, it is quite clear that all these fossils are eukaryotes.

The oldest fossils we find are about 600 million years old and they cannot represent the oldest forms of life, because, for one thing, even the oldest fossils are quite complex in structure and must already have had a long evolutionary history. Furthermore, the Earth is 4.6 billion years old so that the ordinary fossil remains occupy only a little over the final eighth of planetary history, and there was plenty of time for earlier evolution.

Indeed, the fossils we usually study are primarily those multicellular organisms that managed to develop hard structures—shells, bones, teeth—that easily fossilized. Before them must have been multicellular organisms that did not have hard parts, and the earliest of these may have been 800 million years old.

We can go still farther back, however. The American paleontologist Elso Sterrenberg Barghoorn (1915–1984), beginning in 1954, worked with very old rocks in southern Ontario. He shaved thin slices of these rocks and studied them under the microscope. In them, he found circular structures that were about the size of protozoa. What's more, there were signs of smaller structures within these remnants that resembled the kind of structures within cells.

It seemed clear that these were fossils of unicellular organisms, and the oldest of these seem to have been up to 1.4 billion years old. This is nearly twice as old as the o'   multicellular organisms, but even so the history of
       tes still seems to be squeezed into little more

than the final third of Earth's existence. What's more, eukaryotes are sufficiently complex, even in the unicellular form, to require a long evolutionary history.

Sure enough, Barghoorn and his associates detected particularly tiny structures in rocks that were far too old to contain eukaryotes. It now seems that prokaryotes preceded eukaryotes by a long time. The oldest prokaryote remnants so far found have been in rocks that may be up to 3.5 billion years old.

This means that prokaryotes had come into existence by the time Earth was, at most, only 1 billion years old. They then remained the *only* forms of life for over 2 billion years. For all this time they were the ruling life-forms, the true rulers of Earth.

Once eukaryotes arose, it would seem to us that they took over the rulership of the world, first as unicellular plants and animals, then as multicellular plants and animals of various kinds. The predominant organisms of the sea (fish) and those of the land (amphibia, then reptiles, then mammals, and particularly humanity) are all eukaryotes.

However, how do you define "ruling"? The mass of plant life on Earth is ten times that of animal life, and animals can live only as parasites on the plant world. If all plants were to disappear, all animal life would quickly follow them into destruction. If all animals were to disappear, much of the plant world would survive.

To a truly objective extraterrestrial observer, Earth might seem to be a world of plants, with some advanced trees as "rulers" and with an annoying and unnecessary infestation of free-moving parasites. (After all, human beings are composed of trillions of human cells, along with an annoying and unnecessary infestation of parasites on our skins and in our intestines. The parasites are not our rulers just because they live on us.

Let's look at matters in another way. How did eukaryotes develop? There are some who think they arose through the cooperation and eventual amalgamation of prokaryotes of various types.

Thus, prokaryotes that had well-developed genetic mechanisms combined with others that had well-developed systems for handling free oxygen. In combination, primitive eukaryotes developed with a nucleus that was well adapted to genetic functioning and, outside it, mitochondria that were well adapted to handling oxygen. Other portions of the cell also arose from appropriately specialized prokaryotes.

In short, eukaryotes may simply be prokaryote combinations, just as multicellular plants and animals are eukaryote combinations. This point of view is strongly upheld by the American biologist Lynn Margulis (b. 1938).

Therefore, we might view all of Earthly life as falling into three classes: (1) prokaryotes, such as bacteria; (2) combinations of prokaryotes, such as amoebae; and (3) combinations of combinations of prokaryotes, such as human beings.

This may be viewed as analogous to the way in which an American state is a combination of people; and the American federal government is a combination of states (a combination of combinations of people).

A good, efficient, and humane government gives people a much better life than they would have if each lived entirely in isolation and entirely on his own resource, but it is the American view, just the same, that it is the people who are fundamental. After all, people would exist, however savagely and poorly, without government, but government cannot exist without people.

So there is the temptation to say that the prokaryotes still rule Earth.

Let's tackle it from another standpoint. Although eukaryotes came into existence 1.400 billion years ago, and the first multicellular organisms perhaps 800 million years ago, prokaryotes still exist and still flourish.

They exist in such numbers and multiply so rapidly that they evolve at a much faster rate than eukaryotes, either unicellular or multicellular, do. The result is that prokaryotes have evolved into environmental niches that eukaryotes cannot handle. They live at temperatures and at salt concentrations that would kill any eukaryotes. They live on inorganic compounds that could not support other forms of life. As spores, they can survive worse conditions for far longer than any other form of life can. When we develop chemicals to kill them, they gradually adapt to that, and we must find new poisons steadily if we are to control them. They are undefeatable, and when the time comes that some cataclysm, either cosmic or human, destroys life generally, the prokaryotes will be the last to go and may well survive even if all other life vanishes.

Who, then, are the true rulers of Earth, if you think about it without prejudice or self-love?

There is still the question of classifying the bacteria. The first bacteriologists, like Muller and Cohn, tried to divide them up on appearance alone, and we ended up with a collection of names that gave us no notion of how different species are related.

Eventually, as biochemical techniques improved, as bacteriologists learned to study the chemical nature of the cells' constituents, the genes they possess, and the types of chemical reactions they bring about, the chances of exploring prokaryote evolution and working out their relationships improved.

One recently devised system of judging bacterial relationship involves the "ribosomes," which are small objects within all cells, eukaryote and prokaryote alike, that participate in the production of proteins. Since it is the chemical reactions in each cell that give it its distinctive character, and since those chemical reactions depend on the nature of the proteins formed, it would seem that the ribosomes can change only slowly with time. (They don't have much leeway in the types of proteins they can form.) Therefore, the amount of difference in the ribosomes might be a good measure of the evolutionary distance between two species of organisms.

On the basis of ribosomes, it turns out that bacteria fall into two distinct groups. There are the ordinary bacteria that we come across most frequently, whose chemical reactions are much like those of cells in general. They are the "eubacteria" ("true bacteria" in Greek).

There are also bacteria that seem to be quite different in their ribosomes and are therefore, not surprisingly, quite unusual in their chemical reactions and way of life. These are the "archaebacteria" ("old bacteria" in Greek).

The archaebacteria and eubacteria are as different from each other in terms of their ribosome chemistry as either is from the eukaryotes. This means that we can now divide all known organisms that live on Earth into three superkingdoms: Eukaryota, Eubacteria, and Archaebacteria.

Presumably, the archaebacteria are the oldest and most primitive free-living organisms we know of and they include three known subgroups. There are bacteria that cannot use oxygen and have a chemistry that ends in the production of methane, rather than carbon dioxide. They are the "methanogens" ("methane-producers" in Greek). Then there are bacteria that thrive in hot, acid waters and are "thermoacidophiles" ("hot-acid lovers" in Greek). Finally there are those that prefer very salty water and are "halobacteria" ("salt-bacteria" in Greek).

These three known varieties of archaebacteria presumably arose from a common ancestor that remains unknown to us, either because it no longer exists, or because we haven't discovered it yet. One name that I see used for this most primitive of cells is "progenotes" (which may mean "before birth" in Greek).

From these archaebacteria, there may have arisen the eubacteria and the eukaryotes. We don't know if they arose separately from different groups of archaebacteria. One suggestion is that the first eubacteria evolved from the thermoacidophiles, and the first eukaryotes from the methanogens, but I am not ready to believe that.

I think the eubacteria originated first from one group of archaebacteria, but then the eukaryotes developed by combinations of eubacteria. I have no evidence for this. It's just what seems fitting to me.

The eubacteria split into a number of subgroups, of whom a group that contains chlorophyll are particularly interesting. Since the best-known one-celled chlorophyll-containing organisms are the algae, these eubacteria were, for a long time, called "blue-green algae" from their color.

However, they are not algae. Algae are eukaryotes, and the so-called "blue-green algae" are prokaryotes. They belong to different superkingdoms.

For that reason, the "blue-green algae" were first called the "blue-greens," which was a faint-hearted compromise, and then "cyanobacteria" ("blue bacteria" in Greek).

The cyanobacteria may have combined with other eubacteria, forming what are now the "chloroplasts," the chlorophyll-containing structures in plant cells.

The cyanobacteria also produced oxygen during the 2 billion years they may have existed as the only photosynthesizers in the world. It is to them, then, that we owe the establishment of at least the beginnings of the oxygen atmosphere that supports us all.

15
Hot, Cold, and Con Fusion

My dear wife, Janet, is, for some reason, incredibly solicitous over my well-being. If there's a single cloud in the sky, it's umbrella time. If a mist has faintly bedewed the streets, I must slip into my rubbers. If the temperature drops below seventy, on goes my fur hat. I won't even mention the close watch kept on my diet, the inquisitional cross-examination at the slightest cough, and so on.

You may suppose that I am very grateful for all this care. I put it up to any husband in similar straits. "Are you grateful?"

I thought not.

In fact, I complain a great deal about the matter, and I can be very eloquent, too, when I feel aggrieved. And do I get sympathy?

I do not.

To all my complaints, all my friends and acquaintances look at me coldly and say, "But that's because she *loves* you."

214

You have no idea how irritating that is.

So one time recently I was in a limousine being ferried a moderate distance to give a talk. The driver was a foreigner of some sort, who drove with perfect accuracy, and who was clearly intelligent, but he had only a sketchy command of English. Being aware of this, he took the trouble to practice his English on me and I answered carefully and with good enunciation so that he might learn.

At one point, he looked at the smiling sunshine, felt the mild breeze, enjoyed the sight of the nearby park, and said, "It—is—vair byoot'ful—day."

At this, my sense of grievance rose high and I said, in my normal manner of speaking, "Yes, it is. And so why did my wife make me take an umbrella?" And I raised the offending instrument and waved it.

Whereupon the driver, choosing his words carefully, said, "But your—wife—she *lahv* you."

And I sank back defeated. The conspiracy was cross-cultural!

Which, believe it or not, brings me to the subject of this present essay.

Science, too, is cross-cultural, and so are scientific errors. I'm not talking about fraud, now. I'm talking about honest errors by capable scientists. An example I discussed in another volume was the supposed discovery of N-rays in 1903 by a French physicist, René P. Blondlot.

You may be tempted to think that such a thing—overexuberant excitement concerning a startling, but perhaps dubious, discovery—is particularly French. After all, we know the common stereotypical stuff about Gallic volatility and enthusiasm.

Nonsense! Such things happen everywhere.

In 1962, a Soviet physicist, Boris V. Deryagin, reported the existence of "polywater." This seemed to be a new form of water that was found in very thin tubes where the constriction of the environment seemed to compress the water molecules and force them unusually close together. Polywater was reported to be 1.4 times as dense as ordinary water and to boil at 500° C rather than 100° C.

Instantly, chemists all over the world began repeating Deryagin's work, and confirming his results. Perhaps polywater played an important role in the constricted environment of human cells. The excitement was intense.

But then reports filtered out of chemistry laboratories that the properties of polywater would appear if some of the glass of the vessel containing the water was dissolved. Could it be that polywater was actually a solution of sodium-calcium silicate? It turned out to be so, alas, and "polywater" collapsed as thoroughly as N-rays did.

You might say, well, Russians are volatile, too. After all, we know the stereotype of the "mad Russian."

So there's the case of Percival Lowell, an American of the purest Boston Brahman stock (yes, one of *those* Lowells, who speak only to Cabots) and a first-class astronomer.

He reported seeing canals on Mars, and made intricate maps of them. They met at "oases" and sometimes they doubled. Lowell was absolutely convinced they indicated the presence of an advanced technology on the planet, fighting to irrigate the desert midsections with water from the polar ice caps.

Others looked at Mars and saw the canals also, but most astronomers didn't see them. Contrary evidence piled up over the years and we all know *now*, beyond any doubt, thanks to Mars probes, that there are no canals on Mars. Lowell was fooled by an optical illusion.

Does that mean that startling discoveries are *always* wrong? Of course not.

In 1938, the German chemist, Otto Hahn, who had been bombarding uranium with neutrons, came to the conclusion that the results could be explained only by supposing that the uranium atoms broke nearly in half ("uranium fission"). Such a thing had never been heard of and Hahn chose not to risk his reputation by announcing it.

His ex-partner, the Austrian chemist Lise Meitner, however, had been driven out of Germany and into Sweden in 1938 for the crime of being Jewish, and perhaps she felt she had little to lose. She prepared a paper on uranium fission and told her nephew, Otto Frisch, about it, and he told Niels Bohr, who was on his way to attend a scientific meeting in the United States. There he spread the news and the American physicists at once scattered to their laboratories, ran the experiments, and *confirmed* uranium fission. The results we know.

Was that because Hahn was German and Meitner was Austrian? Not at all. I should tell you about the discovery of "masurium" in 1926 by excellent German chemists, but that's for another time.

That brings us to nuclear fusion, which is the opposite of nuclear fission. In fission, a large nucleus breaks apart into two halves. In fusion, two small nuclei join together into a single larger one.

Fission, in a way, is easy. Some large atoms are on the point of dissolution anyway. The short-range strong nuclear force barely reaches across them and their natural vibrations constantly keep them at the point of fissioning. In fact, uranium atoms experience spontaneous fission every once in a long while.

If you add a little energy to the nucleus, fission can take place at once, especially where the atoms are really close to the edge, as is the case with uranium-235. You fire a neutron at it. The uncharged neutron is not repelled by the positively charged nucleus. The neutron slips into the large nucleus and the added instability that results fissions the atom at once.

Fusion is more complicated. Two small nuclei must be

brought very close together if they are to cling to each other and fuse. All nuclei, however, are positively charged and repel each other. Getting them sufficiently close to each other for fusion is an enormous and all but impossible task, it would seem.

Yet fusion takes place in the Universe, and is even extremely common. It will take place in any piece of matter, spontaneously, as long as that piece of matter is (1) mostly hydrogen, and (2) sufficiently massive, about one fifth the mass of the Sun or more.

The nearest place in the Universe where nuclear fusion takes place in massive quantities is at the core of our own Sun.

How does it happen? For one thing, the core of the Sun (or of any ordinary star like the Sun) is at a temperature of millions of degrees. At such temperatures, atoms are broken down and the bare nuclei are exposed. That is important because in ordinary atoms, like those that surround us, electrons are in the outskirts and these electrons act like bumpers that keep the nuclei from approaching each other.

What's more, at the high temperatures at the core of the Sun, the nuclei are moving at enormous speeds, far more rapidly than they can move as part of the ordinary atoms about us. The more rapidly they move, the more energetic they are, and at the temperature of the Sun's core, the nuclei are energetic enough to overcome their mutual repulsions so that they can slam into each other forcefully.

In addition, the huge gravitational field of the Sun causes the outer layers to weigh down upon the core and force those bare nuclei so close together that the density at the core is thousands of times the density of ordinary matter about us.

In very dense matter like that of the core of the Sun, speeding nuclei have less of a chance to miss. Even if they veer away from one nucleus, they may veer right into the next. Consequently both high temperature *and* high density encourage fusion. The higher the one is, the lower the other need be—and the lower the one, the higher the other must be.

In order to bring about fusion, we must have certain combinations of temperature and density and maintain them for a sufficiently long time. The necessary values of temperature, density, and time are known. It is only necessary to attain them all simultaneously.

Since under even the most favorable circumstances, very high temperatures are required, this is called "hot fusion."

Can hot fusion be brought about on Earth? Of course!

We have been doing it for thirty-five years in something called the hydrogen bomb, which is actually a nuclear fusion bomb. It is only necessary to have some fusible substance and produce the necessary temperature and pressure for fusion by use of a nuclear fission bomb (the ordinary A-bomb of Hiroshima) as the trigger.

At this point let me explain that there are three isotopes of hydrogen. There is ordinary hydrogen (hydrogen-1) with a nucleus made up of one proton; deuterium (hydrogen-2) with a nucleus made up of one proton and one neutron; and tritium (hydrogen-3) with a nucleus made up of one proton and two neutrons.

Deuterium is easier to fuse than ordinary hydrogen is, and tritium is easier still. Ordinary hydrogen is by far the most common type, and it is that which fuses at the center of the Sun, but that is too difficult to imitate on Earth. We ought to use tritium, but that is radioactive, breaks down in a period of a few years, and has to be constantly manufactured. Its use as pure fusible material is impractical.

Fusion bombs use deuterium, which is rare but available in quantity just the same. Apparently, a bit of tritium is also added. The fission bomb triggers the tritium fusion with deuterium and that produces enough heat to trigger the more difficult fusion of deuterium with itself. (I don't know the details, obviously, and I don't want to know.)

Recently, it turned out that our nuclear plants producing tritium have been leaking radioactivity into the environment for years. The government, however, anxious to "keep America strong" has kept this secret and let it go on, since to them it didn't matter what happened to the American people as long as America in the abstract was kept strong. (Do you understand that? Frankly, I don't.)

Now the tritium plants are shut down because some busy-body couldn't stand it anymore and made the leaks public. This means our supplies of tritium are slowly dwindling, and when they're gone, we won't be able to explode our H-bombs anymore, unless we build new plants to make tritium, which will, apparently take many years and cost many billions of dollars.

All of this, however, is beside the point. We're not interested in hydrogen bombs, at least not in this essay. The question is whether there is any way of bringing about nuclear fusion in a controlled way, *without* a fusion-bomb trigger, and *without* an all-out devastating explosion.

What we want is to fuse a little bit of hydrogen and use the energy this produces to fuse a little bit more and so on—always just a little bit at a time so that we risk no explosions. Not all the energy will be needed to continue the fusion and the excess we can use for ourselves. Nuclear fusion would be a source of clean energy that would last us for as long as the Earth does.

To do this, we have to get the right combination of density and temperature maintained for the proper length of time. Nothing we can possibly do now or in the foreseeable future will enable us to make the deuterium we're heating very dense, so we have to make up for that by reaching temperatures far in excess of that at the

Sun's core. Instead of something over 100 million degrees, we need something over 100 million degrees.

For nearly forty years, physicists have been trying to obtain the necessary conditions by keeping deuterium gas penned in by strong magnetic fields while the temperature is sent up. Or else, solid deuterium is zapped from a number of sides simultaneously by powerful laser beams that will heat it to the necessary temperature so quickly that the atoms have no time to move away.

But they haven't succeeded. Not yet. Huge devices costing many millions of dollars have not quite gotten deuterium to the fusion ignition point.

Is there any other way of initiating fusion? The crucial point is to get the deuterium nuclei sufficiently close together for a sufficiently long time and then they'll spontaneously fuse and give off energy. The only purpose of very high temperature is to force the nuclei close together against their own mutual repulsion.

But can we trick deuterium nuclei into getting together *without* heat? Can we do it in some ingenious way at room temperature? That would be "cold fusion." Let's consider—

Ordinary deuterium atoms are electrically neutral overall, the negatively charged electron exactly balancing the positively charged proton in the nucleus so that two hydrogen atoms can make contact without trouble. The protons of the two nuclei are then about an atom's diameter (about a hundred millionth of a centimeter) apart.

Every particle has its wave aspects, so that each proton can be considered a wave, with the "particle aspect" anywhere along the wave. (This can't be described properly except with mathematics, but for our purposes we can make use of the wave image.) The likelihood of the particle being at a particular part of the wave depends on the intensity of that part. The center of the wave is most

221

intense and it fades off, or damps, quickly with distance. This means that the proton particle is usually near the center of the wave, though it can be off-center.

In fact, each proton may be far enough away from the center, in each other's direction, so that they may find themselves in actual contact and fuse. (This is called a "tunnel effect," since a particle seems to tunnel through an *apparently* impassable barrier because of its wave properties.) However, when two protons are an atom's diameter apart, the chance of tunneling is so remote that I doubt that a significant number of such fusions have taken place in the entire history of the Universe.

But what if we make the atom smaller? The electron has an associated wave (I'm still using the wave image) and it can only get so close to the proton that a single electron wave circles the proton. The electron cannot get any closer than that. That is the minimum size of a normal hydrogen atom and it just isn't small enough for fusion.

There is, however, a particle called a muon, which is exactly like an electron in every respect but two. One point of difference is the mass. The muon is 207 times as massive as an electron. That means that the wave associated with it is correspondingly shorter than that of an electron. A muon can replace an electron in a hydrogen atom, but because of its shorter wave, it can get much closer to the nucleus. Indeed, "muonic-deuterium" has only one hundredth the diameter of an ordinary atom, and because the muon has exactly the negative charge of an electron, muonic-deuterium is electrically neutral overall and two muonic-deuterium atoms can be in contact without trouble.

Under such circumstances, the two protons are close enough together for the tunneling effect to work easily and fusion can take place at room temperature.

Is there a catch? Of course! The second difference between a muon and an electron is that the muon is not stable. Whereas an electron left to itself would last forever, unchanged, a muon breaks down to an electron and a couple of neutrinos in about a millionth of a second, so there isn't much time for fusion to take place. Muon-catalyzed cold fusion is possible, but totally impractical, barring some unforeseeable breakthrough. Too bad!

Anything else?

Well, hydrogen atoms are the smallest known and they can sometimes sneak into crystals of larger atoms and find homes in the interstices. The champion case of this involves palladium, one of the platinumlike metals. Palladium can absorb nine hundred times its own volume of hydrogen, or deuterium, at room temperature.

The deuterium atoms, by the time they are done flooding into the palladium, are much closer together than they would be in ordinary deuterium gas. What's more, they are held in place quite tightly by the palladium atoms so they can't move around.

The question is whether they are forced *so* closely together that the tunneling effect can become large enough to detect and whether cold fusion may take place at a useful rate. Two chemists thought it worth checking out. They were B. Stanley Pons of the University of Utah and Martin Fleischmann of the University of Southampton, England. They spent five and a half years trying to get cold fusion with simple electrolytic cells that any skillful chemistry student might have set up. They spent a hundred thousand dollars they raised themselves and here is what they did.

They began with a container of heavy water (H_2O with deuterium atoms rather than ordinary hydrogen atoms). They added a bit of lithium to react with the heavy water and create ions that would carry an electric current. They then passed the electric current through the solution with two electrodes stuck into it, one of platinum and one of palladium. The electric current split the heavy water into oxygen and deuterium and the deuterium was absorbed by the palladium. More and more of the water was split, more and more of the deuterium was formed and absorbed by the deuterium, until finally, cold fusion took place.

How did they know cold fusion took place? Well, the palladium electrode developed four times as much heat as was being put into the system. That heat had to come from somewhere, and since it couldn't come from anything else they could think of, they decided it was coming from cold fusion.

Very well. Pons and Fleischmann are legitimate scientists of considerable attainments, and have good track records. They have to be treated with respect.

But—

If anyone finds practical cold fusion and is the first one in the field with it, they are instantly and at once the most famous chemists in the world, a cinch for an immediate Nobel Prize, and if they file for patents, they will become incredibly rich. Naturally, then, Pons and Fleischmann wouldn't be human if they didn't *want*, with unimaginable intensity, to be right. As soon as they had any respectable sign of its presence, wouldn't they decide they had it, even if perhaps they didn't? Human nature is human nature.

Should they have waited until they were *really* sure— till they had overwhelming evidence? For such a startling and unprecedented claim, scientific caution would direct them to wait, but it's easy to advise such caution and very hard to accept it.

After all, nothing would come of it for Pons and Fleischmann personally, unless they were first in the field. There's nothing very outré in the experiment. Scientists knew about this trick of palladium; they understood about tunneling; they grasped muonic catalysis; and they were prepared to set up electrolytic cells. Who could say, then, how many chemists or physicists might quietly be working in the direction of palladium-catalyzed cold fusion. Indeed, Pons and Fleischmann knew for sure there were people at Brigham Young University who were working in that direction.

Apparently, the two groups agreed to send papers, simultaneously, to *Nature*, a very respected journal, on March 24, 1989. However, Pons and Fleischmann apparently could not resist establishing priority. They stole a march by holding a news conference on March 23 and spilling the beans to the press.

This got scientists (physicists especially) furious, for a number of reasons.

1. Giving an important scientific discovery to the press is not the right way to do it. It should be written up as a detailed scientific paper, sent to a scientific journal, submitted to peer review, revised if necessary, and then published. This sounds very academic and roundabout, but it is the only way to keep science on track. What Pons and Fleischmann have done is to put a premium on appealing to the public with work that might be incomplete or ambiguous. If this became the fashion, science would break down in confusion.

2. Pons and Fleischmann did not give full details of the process, which is also the nonscientific thing to do. Naturally, every scientist wanted to test the experiment for himself to see if he could confirm it or find out something else about it (that was what happened in the case of uranium fission). They couldn't be sure of what they were doing, however, because of the incompleteness of the data. Even when Pons and Fleischmann finally submitted a paper to *Nature*, it was so incomplete, *Nature* asked for additional details, and Pons and Fleischmann refused.

3. Pons and Fleischmann apparently didn't run proper controls. They did not describe having done the experiment with ordinary water. Even if deuterium fused under the conditions given, ordinary hydrogen would not, and that should have been tested. If the experiment developed heat with ordinary hydrogen also,

then the source would surely have been something else, not fusion.

4. Pons and Fleischmann reported as their major evidence for fusion the development of heat—but heat can develop from *any* source; it is the common product of all conceivable forms of energy. It isn't enough to say: It can't be this, it can't be that, so it must be fusion. This sort of negative evidence falls down because it might also be some nonfusion process that you just haven't happened to think of or to know of. What is needed is some observation that is positive for fusion and not something that is negative for other things. For instance, if the deuterium atoms fused, they ought to have formed neutrons or tritium or possibly helium-4. This was not reported.

The Brigham Young people did indeed report neutrons but only about one hundred-thousandth as many as those required to produce the heat reported by Pons and Fleischmann. There were so few, in fact, that it would be hard to demonstrate that they weren't derived from neutrons that are always hanging about the environment anyway.

5. Immediately after the announcement, the governor of Utah asked for millions of dollars from the federal government to develop practical fusion in Utah, before the Japanese could steal the notion and develop it in Japan. This lent an unpleasant commercial touch and stressed the economic motivation for haste and incompleteness in the science involved.

6. A gathering of chemists had a field day when Pons and Fleischmann implied they had come to the rescue of the physicists by doing for next to nothing what the physicists had been unable to do for many millions of dollars. There was no need to make fun of honest and rational work. The physicists, being only human, snarled in their turn at the chemists, and what should have been a scientific discussion became a rather unpleasant name-calling piece of scientific pathology.

In any case, I am writing ten weeks after the original announcement. Increasingly, it looks as though the Pons-Fleischmann report will *not* be confirmed; that it will fade away as did the Martian canals, N-rays, and polywater. Too bad, for the world could surely use practical cold fusion.

But to end on a positive note: All the returns aren't in yet. There is still an outside chance, a very outside chance, that cold fusion will be confirmed. By the time this essay appears, we should know absolutely, one way or the other.*

And second, even if Pons-Fleischmann turns out to be a chimera and a mirage, an enormous effort is being put into investigating electrolytic cells with palladium electrodes and who knows what people may find out as a result. Something interesting perhaps. Perhaps even cold fusion by some other route. I certainly hope so, though I must admit I wouldn't bet on it even with generous odds in my favor.

*We do! Cold fusion is as dead as a doornail.

16
Business as Usual

I love giving talks, but I can't give as many as I would like. For one thing, the business of my life is writing, and I cannot sacrifice too much writing time for the pleasure of talking—even at the high lecture fees I routinely extort. In the second place, I don't like to travel, so I don't accept engagements that are more than a few hours' distance from my home, especially in the winter months.

But if the talk is close to home and comes at a time when I can spare an evening, I refuse to turn it down simply because I don't know anything about the subject. After all, I can always brush up on it quickly, and since I like to think of myself as a person of infinite resource and sagacity, I am always sure that I can think of some approach I can handle.

Thus, it came about earlier this year that I was asked to give a talk of the future of "smart cards."

I drew a complete blank. Smart cards?

However, the fee was right, the place was right, the time was right, so I had no intention of refusing. I wrote the people a letter and said, "I'll be glad to oblige, but tell me—What are smart cards? Cards that play poker by themselves?"

I was promptly deluged with material on the subject. Smart cards are objects very much the size and shape of credit cards, but are so thoroughly computerized that they carry enormous amounts of information about you and your affairs and can greatly simplify the financial transactions you might wish to undertake.

That was a relief. After all, I had written a brief essay on the subject back in 1975, strictly out of my head. I just didn't know the things were called "smart cards." So I gave the talk with confidence, and it was very successful, I'm glad to say.

In the process, though, I had to do some thinking along lines I had not previously dealt with much, and I would like to share some of those thoughts with you, for what they're worth.

Suppose we start by considering at what point in history human beings became human beings.

From the biological standpoint, that may have come about as early as 5 million years ago, with the appearance of the first australopithecines. They were the first organisms to qualify as "hominids"—that is, as creatures that more closely resemble modern human beings than they resemble any apes, living or extinct.

We might wonder why we should give the australopithecines the honor of human relationship (if honor it be), when they were somewhat smaller than modern chimpanzees, with brains no larger and with a life-style not much different.

There was, however, one important difference between australopithecines and apes that made up for everything. The australopithecine spine was bent backward in the lumbar region, as ours is, and from that and from what the surviving hip and thigh bones tell us, we are quite certain that the australopithecines walked erect as easily and as comfortably as we do.

Apes, bears, and other creatures can get up on their hind legs but it is an uncomfortable and temporary stance for them. With hominids, it is permanent. Furthermore, unlike such bipeds as birds, hominids have saved their forelimbs from over-specialization and have made them into manipulative appendages of marvelous delicacy. The upright stance and the possession of well-developed hands with opposable thumbs separate hominids sharply from all other Earthly forms of life, living or extinct.

That's all very well, but every species has its own adaptations, some of them remarkable indeed. Viewed by an impartial observer, the adaptations of many nonhominids might seem more noteworthy than the hominid spines and hands.

When, then, did the hominid way of life develop to the point where the difference between hominids and nonhominids became an astonishing cultural gulf and not merely a matter of biological minutiae?

The easy answer to that and the one I've always given has been that the difference arose with the use of fire.

As far as we know, the use of fire is at least 500,000 years old, and the first to use it were members of *Homo erectus*, a relatively small-brained ancestor of ours.

The use of fire makes the distinction satisfactorily. No group of modern human beings (*Homo sapiens*) has, as far as we know, been without the use of fire. On the other hand, no other species of living thing, alive or extinct, other than *Homo erectus* uses or has ever used fire, or had anything to do with it except to run from it if it could.

Can we go farther back, however? Is there anything prior to the use of fire that we can use for the hominid-nonhominid distinction?

To be sure, human beings use tools, but so do other animals. Chimpanzees can use sticks. Beavers build dams, spiders build webs, birds build nests, and all of these things might be viewed as tools.

Even the most primitive hominid tools were probably more elaborate than any nonhominid tools, I'm sure, and we might try to make a distinction between instinct and reason, but this is shaky ground. As far as the use of tools is concerned, we may be dealing with differences in degree rather than in kind.

Even if we try to define the hominid attribute as tool making rather than tool using, it might still be a matter of degree. Beavers, to some extent, have to shape the logs they use. Chimpanzees strip leaves from a twig before using it to catch termites, and so on.

About 2 million years ago, however, the first organisms had evolved who were hominids sufficiently like ourselves to qualify for placement in the genus *Homo*. They were *Homo habilis*, and with them came the development of stone tools. More than that, they apparently *shaped* the stone tools in very primitive fashion.

Now sea otters use stones on which they break the shell of the mollusks they feed on, but they use unmodified stones. *Homo habilis* was the first species ever to *modify* this resistant material. It may be, then, that stoneworking is the first cultural or technological development that draws a firm line between hominids and all other forms of life.

At this point, inspiration struck me and I came up with an idea that, as far as I know, is original with me.

Prior to stoneworking, hominids surely used tools and these consisted chiefly of bone and wood, each of which is far easier to modify than stone is. Such tools, however, presented no problem. They were available to all. Anyone could obtain a long bone by chewing away the meat around it, and anyone could break a branch from a tree. There is no particular skill involved in chewing or breaking.

231

When the time came, however, that hominids were deliberately manufacturing bits of rock with sharp edges or points, objects that could then serve various useful purposes, the question of "skill" arose. After all, some people were bound to be more skillful at stoneworking than others, which meant that some people were bound to have more and better stone tools than others.

Now if one hominid had a bone or a tree branch and another one did not, the deprived one could always get one of his own that was just as good. Bones grew in animals and branches grew on trees.

If one hominid, on the other hand, had a good stone tool and another did not, and if the deprived one lacked the skill to make a satisfactory one for himself, what could be done about it? The deprived one might try to take the other's tool by force, or to steal it when the owner wasn't looking. Either way, there was going to be a fight and a lot of hard feeling and bruises.

There must have come a day, then, when a man without a stone tool managed to get hold of some food and must have said to a man with tools, "Look, you have five cutters. You don't need all five. Give me one of them, and I will let you have all this food which I have worked hard collecting and which you will be able to eat without having had to do any of the work." (Naturally, we assume he said all this in whatever kind of grunts and gestures *Homo habilis* used.)

When, where, and how this first happened, we don't know and can't tell, but we can be sure that at some time well before the discovery of fire, hominids had begun the practice of barter, and this was the invention of "business," so to speak.

Barter, even in its simplest form, is a tremendous advance. It tends to even out the resources available to a group of people. By trading, each person can make use of not only his own skills, his own work, and his own luck, but also the skills, work, and luck of others. Each person involved in bartering gives something he has little

232

need for, in exchange for something he has much need for, so everyone gains.

This is something completely new in the world of life. A male bird feeds a female bird incubating her eggs, a mother protects its young, one monkey grooms another, but these are inflexible behavior patterns, and seems qualitatively different from barter.

The deliberate use of barter to increase the general standard of living, involving, as it does, reason and judgment, is a strictly hominid invention and it may possibly be nearly 2 million years old. The art of stoneworking and, even more so, the development of business would thus seem to be the moment when hominids became something special and unprecedented in the history of life.

The invention of business meant that there were now two forms of interaction when two groups of hominids happened to cross each other's path. They might fight, each trying to establish ownership of the territory on at least a temporary basis. Or they might do business, since each side might possess something the other would covet.

Undoubtedly, fighting was the more traditional and the likelier of the two alternatives, but as technology developed and as individual groups of hominids found themselves with more valuable objects in greater variety, the chances of business increased.

We might generalize and say that throughout the development of culture and civilization, there was always the choice of destructive or constructive interaction between social groups—war or trade—and that, on the whole, the constructive interaction must have won out since culture and civilization have grown continually more elaborate and versatile—at least up to now.

The gradual (and perhaps heartbreakingly slow) predominance of the constructive alternative not only allowed technological advance, but actually made it inevitable. The needs of trade encouraged travel and a broadening of horizons, for the greater the area over which barter could take place, the larger the pool of resources that could be spread out and evened, and the higher the standard of living that resulted. I don't say that people *deliberately* brought this about through far-sightedness; it was just the natural result of continuing to seek self-benefit through trade.

Why would anyone develop rafts, for instance, and then boats? To go sailing on a river and have a good time? Why would people build roads and develop wheeled carts pulled by donkeys? To bet on races?

No, boats and carts were devised only for the purpose of traveling up and down rivers and roads, to trade and transport goods. Even in the very early days of civilization, bits of amber from the Baltic seacoast found their way to the Mediterranean and pottery from different areas scrambled themselves widely up to the limits of the relatively advanced areas and even beyond.

As civilization advanced, those who benefited from its technology and enforced discipline learned to use that to fight off the uncivilized hordes outside, who preferred to take rather than trade. And even when the "barbarians" won out and enforced a breakdown in business transactions that brought on a decline in the standard of living, they themselves quickly learned that this was not a desirable situation. With surprising quickness, they adopted the civilization of those they had conquered. In this way, win or lose, civilization spread outward until it covered the world.

However, civilization could not advance far on the basis of barter alone. Barter had the effect of raising the standard of living generally, but it had its disadvantages, too. (Everything does!)

The objects being bartered are digital, for instance, and involve things that are not easily divisible or comparable. Is a goat worth three chickens or four chickens? Maybe it's worth three and a half but a half-chicken doesn't lay eggs or reproduce. And if barter involves pots or sickles, for instance, then half a pot or half a sickle is totally useless.

It is easy, then, perhaps even inevitable, that in many barter arrangements, both sides end up feeling cheated. That must be why, in the ancient Greek myths, the god Hermes is both the god of merchants and the god of thieves. When I was young and totally innocent, that puzzled me. Now that I am old and slightly guilty, it occurs to me that the clever Greeks failed to see any distinction between the two lines of endeavor. In modern times, for instance, we might invent the great god Whereas who might be the god of lawyers and also the god of connivers. Or the great god Sniff-Sniff, who would be the god of critics and also the god of fools.

But I digress.

Fortunately, by about 3000 B.C., metals were well known in the Middle East. There were gold, silver, and copper, all of which could be found free as metallic nuggets, or could be easily obtained from appropriate ores. There was also the copper-tin alloy called bronze. These metals were much desired. Bronze was hard enough to be superior to stone as a material for tools and weapons. Copper, silver, and gold—particularly gold—were highly ornamental, and could easily be worked into a variety of fascinating shapes. The desire for ornamentation is deeply ingrained in the human psyche so the possession of metals, particularly gold, was coveted.

Gold is beautiful beyond question. It is very rare so that getting even a small amount is unusual and is a cause for great self-congratulation. In addition, it is indestructible if left to itself. It doesn't rust, melt, fade, or lose its luster in the slightest.

What's more, gold is not digital, but analog. It can come in all kinds of weights. If a piece of gold is broken in half, each half continues to have half the original value.

It would be sensible, then, to barter a tiny bit of gold for a large ox, let us say. Not only is the gold much more portable, but its weight can be slightly adjusted up or down, so you don't have to decide whether it is worth three whole goats or four.

Although gold has no value in itself, aside from its use in ornamentation, it is of infinite value as a way of facilitating trade. There is no question that the invention of a "medium of exchange," of "money," had a catalytic effect on business. The existence of gold and its movement back and forth, therefore, raised the standard of living and inevitably brought about changes that advanced the level of culture and civilization.

But the gold had to be *used*. Thus, the pharaohs of Egypt had themselves buried with all sorts of golden treasures and did so with the greatest precautions against tomb robbers. However, every last pharaoh's tomb was quickly robbed, even that of Khufu at the center of the monstrous "Great Pyramid." (The tomb of Tutankhamen escaped by a fluke, but never mind that.)

When I was young and innocent, I felt indignation over the tomb robbers, but as I grew older, I recognized them as the saviors of civilization. The removal of all that gold from circulation would have devastated the economy of Egypt and the rest of the ancient world. The tomb robbers, by restoring the gold to circulation, performed the acts of heroes, for I need not tell you what happened to them if they got caught.

Of course, there were still inconveniences. Merchants and traders had to carry around balances to weigh the gold. The balances had to be honest ones, with arms of equal length. The standard weights used had to be of correct value. The use of dishonest weights or balances would once again blur the distinction between a merchant and a thief. What does the Bible say? "A false balance is abomination to the Lord: but a just weight is his delight" (Proverbs 11:1).

In the eighth century B.C. came another important

236

invention. The kingdom of Lydia, in western Asia Minor, placed the matter under governmental regulation. They issued small disks of gold of guaranteed weight and stamped each disk with its weight and with some royal symbol to indicate its honesty. In short, they invented "coins."

Away went the balance and, to an extent, the suspicions of crooked weights and measures. Once again trade and business were encouraged and prosperity grew. Lydia benefited enormously by the invention and its last king, Croesus, grew so wealthy that "as rich as Croesus" is still a common catch-phrase. Naturally, the use of coins spread rapidly throughout the civilized world.

Governments can themselves be dishonest. There is always the temptation for a ruler to try to save money by adding less valuable metals to the gold coins and for paying off his debts with coins worth less than the face value.

This sort of thing would, however, invariably boomerang. As people grew reluctant to accept the cheaper coins, trade languished, business fell off, the economy suffered, and living standards declined. The government was then forced to debase the coinage further, making things ever worse.

On the other hand, those nations that managed to keep up an honest coinage had their coins greatly desired for the purposes of trade and business. They then remained economically strong, despite the vicissitudes of war. The "bezants" put out by Constantinople and the "ducats" put out by Venice are an example of honest coins put out by nations that remained prosperous over long periods of time.

There is some inconvenience in the fact that gold has an intrinsic value. Even if the government issues honest coins of honest weight, individuals might clip off tiny portions along the edges and save them. Eventually, they would accumulate enough gold shavings to represent a consequential sum—stolen from the people, generally. Governments had to mill coins, placing tiny ridges along their circuit, so that clipping a coin would be too obvious to get away with.

In general, every advance leads to ingenious methods of stealing, which are countered by ingenious defenses, met by more ingenious crookedness, and so on indefinitely. Again, on the whole, despite our cynical tendency to believe otherwise, honesty wins out, or civilization and culture would not have advanced as they have.

In the Middle Ages, the Church's strictures on usury (lending money at interest) served to keep the economy in ruins after the fall of the west Roman Empire. Without usury, there would be no loans, and business would be choked down with the usual bad effects that followed. The Jews, who were not affected by Christian views, kept the economy creeping along through their work as moneylenders (the only gainful employment allowed them by pious Christians) and their reward for this was to be viewed as Shylocks.

With the Renaissance, the Italians were finally forced to choose between business and holiness and they chose the former. They established banks which lent money at interest, and which became connected with each other by letters of credit. The Italian business tycoons then had money enough to patronize art, and that kept the Renaissance going.

China, during the Middle Ages, had invented the use of paper money (not the metal stuff that people were used to, but promises to pay the metal stuff on demand). That gradually spread to Europe.

Paper money further increased the ease of doing business, at least as long as people trusted the ability of the government to redeem the paper with coinage on demand. However, while there was a finite amount of gold, there was an essentially infinite amount of paper. Governments could almost never resist printing up endless reams of paper bills with which to pay their debts on the assumption that the population would never demand redemption in great numbers all at the same time.

However, as the paper money supply rises, there is invariably an increasing reluctance to accept the increasingly worthless promises of redemption. More and more of the paper is demanded before a sale can be made (in the hope that with *enough* paper, a fair amount of coinage can be obtained). In other words, there is inflation, which can reach catastrophic proportions.

And there is also the problem of counterfeiting, which is fought by making the bills ever more complicated, which in turn forces the counterfeiters to ever higher flights of ingenuity.

Refinements have continued right down to the present. It has become common for each individual to have his own supply of paper money, not in fixed denominations but in any amount he wishes. The paper money is in the form of checks, made out to the penny, which become valid only on signature. It is clearly easier to write out a check at need than to walk about with a huge wad of bills.

Of course, the check is, once again, merely a promise to pay in "real" money on demand, and people hesitate to accept a check from a stranger, since he or she might not have the bank account to redeem it with. (And there's also the problem of forgery.)

Things became still simpler with credit cards, where one's ability to redeem the amount is more easily checked, and where one ends up having to write only one check a month.

Now we have smart cards.

In fact, we no longer have to exchange money physically, not in the form of cows and chickens, nor of gold and silver, nor even of paper money and checks. We now have the capacity to carry through transactions electronically, by making controlled changes in symbols. The flow of business has become easier than ever.

Since electronic transactions can be carried out at the speed of light, the whole world has become a single business unit in which any transaction anywhere by people separated the full width of the planet can be carried through in seconds.

What started as simple primitive barter, then, a device that spread the resources of a few individuals to the benefit of those few, can now (at least potentially) be a device for spreading the resources of all the world to the benefit of all the world.

Does this mean we are now living in a Utopia? We should be, but there are still problems. Even if we place to one side (for now) the physical problems such as overpopulation and overpollution, we find enough trouble just in the act of money handling.

The less-developed portions of the world can, for instance, borrow money easily (albeit at high rates of interest) on the supposition that the money will be used to develop the resources and economic structure of the nation, thus enriching it and enabling it to pay off the loan and to improve its standard of living as well. Instead, the money is all too often diverted, through corruption and incompetence, into the pockets of a few, leaving the land poor and with a debt load it cannot possibly repay.

Then, too, as the economic structure of the world becomes more complex, there is always the temptation for unscrupulous individuals to take advantage of their positions of power and influence to enrich themselves at the expense of the public generally. This means we have influence peddling in Washington and insider trading on Wall Street.

That leaves us now with the matter of the future. Will the destructive influences finally win out over the constructive ones and will civilization and culture at last collapse?

At present, we can only wonder.

17
Just Say "No"?

Some weeks ago, I was attending a function in New York and a friend of mine from the sticks was present. He had once lived in New York City but was now living at a place I shall call "Sleepy Hollow," for that is not its name.

My friend favored me with a long tirade on the nastiness and unpleasantness of New York City—mentioning its noise, its dirt, its crowds—and contrasted it with the bucolic charm and rustic delights of Sleepy Hollow.

I listened with pained patience. I am used to people from outside the city who come to the city (in order to do *something* with their lives, since the chief intellectual activity in Sleepy Hollow is collecting a tan) and then throwing scorn upon it.

Afterward, though, I thought: Why *should* I listen with pained patience and endure the insults? Why don't I answer with something like the following—

"See here. You are in my city. You are speaking to a person who was brought up in this city, lives here, and loves it. You are speaking to one who finds it, despite its faults and problems, the most stimulating environment ever invented by humanity. My life is filled with variety and adventure because I live in this city, for just walking its crowded, noisy, dirty, smelly streets introduces me to a microcosm of the world, and to all its peoples and cultures. No one has asked you to come to New York City if you don't like it. You are welcome to stay in Sleepy Hollow where you can satisfy your need for excitement by indulging in deep-breathing exercises. But if you do come here because every once in a while you want to experience life, please keep a civil tongue in your head and don't insult my city."

I don't know if I can actually ever bring myself to say something like that. It would be impolite. It would also be impolite for me to visit Sleepy Hollow and say to its inhabitants, "What do you [yawn] do around here? Is there any place to [yawn] go, or [yawn] anything amusing and interesting? Whom do you speak to after you've exhausted your three [yawn] neighbors?"

The trouble is that New Yorkers are polite people who don't do things like that. We leave it to out-of-towners.

And then, this last weekend I had my chance. I was at a resort hotel upstate (yes, I leave Manhattan now and then for short distances—under duress) and listened to a lecture on the New York City water system. It was an excellent address, but the speaker somehow got the idea that he was addressing a rustic, upstate crowd.

He explained that he had been born and brought up in New York City but that "in 1970, I was privileged to leave the city for these wonderful country surroundings."

He was just stroking what he conceived to be the audience, but I stiffened in my front-row seat. I waited for the question period, but it turned out, I didn't have to. Somewhere in the course of his speech, it occurred to him to wonder whether anyone in the audience was actually from New York City. "How many of you live in New York City?" he asked.

Half the hands shot up, mine among them, of course, and my voice rang out in unmistakable hostility, "We haven't been *privileged* to leave our city, sir."

I'll give him credit. He caught the faux pas at once and apologized.

But it got me to thinking about a number of things, and if you don't mind (or even if you do), I'm going to devote this essay to nonscientific comments and express my views on certain social phenomena. You may not see that what I am about to say has anything to do with insulting New York City, but I promise you that I will tie it all together before I am through.

On June 3, 1972, I was receiving an honorary degree of Doctor of Letters from Alfred University in south-central New York State. On the program with me, and delivering the commencement address, was Rod Serling, who was getting a degree of Doctor of Humane Letters. (Mine were inhumane, I suppose.)

It was the second time I had met him, and I was terribly pleased for I was a great admirer of his "Twilight Zone" television series.

Now suppose that, as we were both sitting there on the dais during the commencement exercises, some troublemaking demon had whispered in my ear, "In three years, one of the two of you, Serling or yourself, will be dead."

I'm afraid I would have gone into a decline, for the odds that death would tap me rather than Serling would have seemed to me to be very high. To begin with, Serling was five years younger than I was. Second, Serling was slim and highly tanned. Undoubtedly, he took care of himself physically, working out, eating abstemiously, and so on. I, in contrast, was flabby, plump, and pale. I lived an absolutely sedentary life in front of my typewriter, and when I fell out of bed in the morning, I felt that to be sufficient exercise for the day.

To be sure, Serling was a hard-driving man with a life that was full of deadlines, and of goals that had to be met—but my life was exactly that, too. In that respect, we were even.

And yet on June 29, 1975, Rod Serling died during open-heart surgery at the age of fifty—and I am still alive.

Why?

The answer is straightforward. Serling was a chain-smoker from adolescence and the smoking had turned his circulatory system into a set of breakable clay pipes. Everything he did to preserve his health, and everything I neglected to do to preserve mine,* did not matter compared with the fact that he had smoked perpetually and I did not smoke at all.

Tobacco, after all, is an addictive drug and it does harm. It kills, by its direct and deleterious effect on the body, hundreds of thousands of Americans each year. It is the greatest cause of easily preventable death in the world.

Even before it kills (and the killing process is a slow one), tobacco smokers are, on the average, more often ill than non-smokers are, have more respiratory problems, lose more workdays, perform their work less efficiently when they do work, and overload the medical system of the United States unnecessarily.

Worse than all that, tobacco is one drug that directly affects more than the practitioner. Alcohol remains within the drinker's own bodies, and so do other drugs. Tobacco smoke, however, after permeating the smokers' lungs, is released into the air for other people, who may be nonsmokers, to breathe.

It doesn't matter whether you draw the noxious smoke into your own lungs voluntarily, or breathe it involun-

---

*Let my readers rest easy. Since my own mild heart attack, I have reformed. I have lost weight; I exercise a bit; I eat with discretion; and so on.

tarily after it has emerged from someone else's lungs. In either case, it is toxic. A smoker may know the risk and prefer to give up, let us say, five years of life for the fifty years of pleasure smoking might give. That's his business, and I agree with his right to accept such risks.

But why should nonsmokers be exposed? Each breath of smoke shortens life and why should nonsmokers give up any part of their life for the sake of preserving the smoking pleasures of others?

(I do not mention the way in which the smoky effluvium clings to the clothes of nonsmokers, and how smoke in restaurants ruins the taste of food for nonsmokers. Smokers may, for all I know, enjoy stinking, and may get pleasure out of eating steak that tastes like old ashtrays, but not so the rest of us. Nor do I mention the number of fires, whether in forests, hotels, or homes, caused by smoldering cigarettes.)

Fortunately, the situation with regard to smoking is now well understood. Nonsmokers are fighting vigorously for their right to smoke-free air, and are winning. I'm against the prohibition of smoking, of course. Everyone has a right to the death of his choice but only in places where that death is not afflicted on others.

That brings me to alcohol. Smoking is only four centuries old where Europeans and their descendants are concerned, but alcohol consumption dates back to prehistoric days, and distilled liquors are about six centuries old.

I won't bother preaching on the harm alcohol does. There is probably no one in the United States who hasn't had some experience with alcoholics. Nor do I suggest we ought to take violent measures to prevent alcohol consumption. That was tried in the 1920s and proved a dismal failure.

However, ought we to have a society that openly encourages drinking, that, in fact, *insists* upon it?

I am constantly plagued with invitations to attend cocktail parties, for instance, and I am sometimes forced to attend them because it may be one that requires my presence, or because it is being held in honor of someone or something I cannot turn my back on, or because it is hosted by people or institutions to which I owe gratitude or loyalty.

At these cocktail parties, you are *expected* to drink. If I stand there with an empty hand, I am bombarded with an anxious chorus of "Can I get you something to drink?" What I usually do is to order a ginger ale, which has the virtue of looking like something alcoholic, and sipping it slowly. If I can have a maraschino cherry thrown into it, all the better.

I remember one cocktail-party about twenty-five years ago, at which I was caught without a ginger ale in my hand. A woman approached me with a cocktail in one hand and a burning cigarette in the other. I automatically leaned away from her to avoid the smoke as much as I could.

I suppose she noticed that and, having already consumed a number of cocktails, said to me belligerently, "What's the matter? Don't you smoke?"

"No, I don't," I answered briefly.

She noted my empty hand. "I suppose you don't drink either."

"No, I don't," I answered.

Whereupon she said angrily, "Then what the hell *do* you do?"

And in a normal speaking voice, I answered, "I f—k a lot."

That ended that conversation.

The point I'm making is that "social drinking" is encouraged as part of the niceties of life. I remember a couple on television talking about their return to nature—about living on a farm and indulging in the simple life. However, they explained, that didn't mean they weren't sophisticated. They "always had wine with dinner."

Well, what's wrong with social drinking?

Just this: We live in a complicated technological world, and every one of us is constantly at risk. We depend on others for the workings of those devices, machines, and systems that supply us with the basic necessities of life, and we count on good judgment and efficiency.

When a ship strikes a reef and pours out millions of gallons of oil into the sea; when a train strikes a truck and is derailed; when a plane crashes shortly after take-off; when any of a thousand and one things go wrong—the fault may lie in circumstance, in untoward events, in the weather, in equipment failure, and so on.

On the other hand, it might also be the result of some personal factor. Someone made the wrong decision, re-acted inappropriately or too slowly, missed something that should not have been missed.

Such things happen even to perfectly functioning human beings under the best of conditions, but they happen *more often* to drinkers. After all, you don't have to be dead drunk to bring about catastrophe. A social glass is enough. One or two little "drinky-poos" may be suffi-cient to dull one's senses, slow one's reflexes, dampen one's thinking mechanism, to where one is not quite on the ball.

By actually *encouraging* people to swim in an alcoholic haze, our society is asking for all sorts of disasters.

Let me interrupt myself just a bit in order to avoid play-ing a holier-than-thou role. For a variety of reasons, I have never learned to smoke and drink, but I have my drug, too. It is called "writing."

It has all the qualities of a drug. It is addictive. I can't stop writing. If circumstances force me away from my typewriter, I quickly develop withdrawal symptoms. While I am writing, I experience a "high." The cares of the day vanish and life becomes a lark. You might even argue that my writing, like smoking and drinking, has a deleterious effect on my health, for it forces me into a sedentary life that has a tendency to make me fat and flabby, and to underutilize my muscles.

It even causes me to neglect alternate pleasures. Despite what I said to the woman at the cocktail party, I am not an inveterate sex addict. I have the impulse to be one, but as I have frequently said, "You can be a Don Juan, or you can publish 459 books; you can't do both." Well, long ago, I made my choice. Someone once asked me, "If you had your choice, Dr. Asimov, would it be women or writing?" My answer was, "Well, I can write for twelve hours at a time without getting tired."

Yet there is a difference. If I were to smoke or drink, I would in neither case serve humanity in any way by indulgence in my vice. If, in sudden enthusiasm, I increased my output of cigarette ash or of empty cocktail glasses, no one in the world would be benefited.

I like to think, however, that my writing addiction is useful; that what it produces gives pleasure to other people. Surely, that's something.

Now I'm ready to go back to my friend from Sleepy Hollow.

As I write this, there is a great surge of East Germans working their way into West Germany. Presumably, they are moving in the direction of greater freedom. They are also moving in the direction of a higher standard of living, and I must confess that I'm not sure which is the true driving force.

The reason I'm not sure is that there is a similar movement in the United States, where it is obvious that the motion is not in the direction of greater freedom but in that of what is seen as a better life. The middle classes are leaving the older cities and are flooding into the suburbs and exurbs.

The escapees in this case are not in search of a right to vote; they're not fleeing the all-pervasive eyes of the secret police. They're simply heading for swimming pools, for manicured lawns, for good schools and safe neighborhoods, for golf courses.

It is difficult to blame them, but consider that the flight from the city is the flight of the middle class, the substantial citizens, the tax-paying elite. Into the city flock the poor and dispossessed in search of whatever fragments the city can afford them.

The result is that from year to year, the population of the older cities includes a larger percentage of the poor and dispossessed, who do not, and cannot, pay substantial taxes, but who, in fact, require welfare and help of various sorts. The population also includes a steadily smaller percentage of the settled, better-off element.

As a result, the tax base of the cities erodes, the cities become shabbier, and decay more rapidly, and those people who have remained but who can afford to leave are ever more anxious to do so. The change for the worse thus accelerates.

Now, then, do the suburbanites worry about this? I presume that many do, but I also presume that many don't. I think that the general mood of the country is that cities are festering sores the nation would be better off without. I am even cynical enough to get the feeling that some of the escapees are delighted the cities are doing so badly and, far from wanting to help them, are eager to shove them further into the mire.

Why should this be? Well, when I was young I would occasionally go to some summer resort in the mountains for a week or two in the heat of midsummer. I couldn't help noticing that whenever a new batch of city people would arrive, everyone who was already there would ask eagerly, "How is it in the city?"

I decided I knew why the question was asked. I checked it out and I was right. Whenever I came in and was asked the question, I would answer (regardless of the true facts), "The weather in the city was wonderful. It was mild and dry. I hated to leave."

I would then watch their faces fall and their lips being chewed. They were spending money on a vacation and there was no value in it unless it was clear that those who were *not* on vacation were being tortured by horrible weather.

It was like the old Hollywood slogan: "It's not enough to succeed; your friends have to fail."

Of course, when other newcomers would tell the truth about the city and discuss the heat and humidity, the vacationers would perk right up and smile and revel in the misery of the millions.

It's human nature. Some would call it bestial, but it isn't, for animals are not like that. It's *human* nature.

So it is not enough for the Sleepy Hollow people to enjoy their lawns and trees and quiet and fresh air. They must come into New York and talk about the dirt and the noise and the smells and the danger. Every bit of these bad things make them feel smarter about escaping. The greater the contrast, the better off they are. So they don't care if the city is going to hell. The faster, the better. They'll enjoy their suntans more.

Now let's get back to the city. The East Germans can escape to West Germany with its greater freedom and higher standard of living. Poor people all over the world, hungry people, frightened people, weary people, hopeless people, have a dream. Someday they may be able to find their way to the United States—the richest country in the world—the most advanced—the freest.

This is no illusion, no useless dream. My parents and I, many years ago, came to this country as penniless immigrants and I lived the American dream and became a substantial citizen.

So it is that across our borders come people from Mexico, from Central America, from southeast Asia, from Eastern Europe, from everywhere, all with the dream in their heart.

Now tell me, where do the poor and downtrodden, the homeless, the hungry, the sick, who are right here

251

in the great American cities (and some of whom have been Americans for generations)—where do *they* go and what is *their* dream? There is no other country they can dream of going to for they are already in the country of everyone's dreams. Only they haven't made it, and most of them know they're not going to make it.

Nor can they do as other American cityfolk do and escape to the beautiful neighborhoods of the suburbs and the surrounding countryside. They haven't got the money for it, and to tell the truth, the people who have already escaped are not anxious to have the dregs of humanity follow them. Yonkers nearly went bankrupt rather than agree to build low-cost housing among the affluent.

What's more, the poor in other countries know that they are living in poor countries. They know that not only they but just about everyone is on the edge of starvation. The poor in the United States, on the other hand, are daily told on all sides that the United States is rich and wonderful. Movies and television constantly show them what happy lives other people are leading. The poor and the hopeless are made to feel alone and imprisoned by misery in the midst of a victorious and wealthy and endlessly self-congratulating society.

So what do the poor do, short of rising in rebellion and being shot down by the forces of society? They've got to feel better *somehow*.

They turn to drugs. Why not? All the yuppies have their cigarettes and cocktails despite their good life, and the poor smoke and drink what they can, too. But the poor need something that will work faster and kick them up higher and that means heroin, cocaine, and crack.

Now the people of the United States have decided that the drug epidemic is the greatest danger facing the country, especially since it is spreading to the suburbs, and a great many of the beautiful people are snorting the occasional social cocaine along with their social cocktails.

But what is the cause of the epidemic? Well, *partly* it's the decline of the cities as the middle class moves out and proceeds to disown the city and tell it (as ex-President Ford was once supposed to have told New York City) to "drop dead."

The deeper cause, of course, is the imbalance of our society. The Reagan administration felt that the best thing to do was to help those who were well-off—to lower their taxes and turn a blind eye to their unethical practices—in the hope that the endless money they accumulated would trickle down to the poor and improve their lot.

That made for a lot of popularity for Reagan, but it didn't work, just the same. The rich indeed grew richer under Reagan (I did, too) but the poor have grown poorer and more numerous.

And how do you treat the drug addiction that is forced on people by sheer hopelessness? Nancy Reagan had a solution: "Just say 'No'!" It's my opinion that Nancy couldn't possibly say "No!" to a new dress or to a new astrologer, but she expected miserable humanity to say "No!" to the only thing that made them feel good.

What else? Put extreme pressure on Colombia to suppress its drug lords? Send them some helicopters and tell them to endure the bombings and the assassinations by outlaws who have enough money to buy up most of the police force and the politicians and murder the rest? Do you think they will really do this?

For goodness' sake, we banned the production and sale of alcoholic beverages in 1920, and the United States was at once converted into a gangster-ridden society, with gang wars and endless crime. Did we fight it? Did we handle it? Yes, indeed. You know how? We gave up and, in 1933, made alcoholic beverages legal again.

Do we really expect Colombia to fight a war we couldn't fight, when Colombia is so much weaker than we, and their criminals so much stronger than ours? Surely, you jest.

So what to do? More force? More jails? More police power? Army involvement?

One, it might not work. It didn't during prohibition.

Two, it would be expensive, and since Reagan spent a trillion dollars building up the armed forces, and more than doubled the national debt to avoid taxes, and saddled us with a heavy unfavorable balance of trade to avoid inflation, the nation has lost its desire to spend money on *anything*.

Three, even if it were done and if it worked, it would mean an erosion of American liberties. We would become accustomed to police searches without warrants, to arrests on suspicion, to limitations of movement, etc. etc. And once lost, liberties are very hard to regain.

Well, then, what would *I* do? For one thing, I'd like to end some of our hypocrisy. Tobacco and alcohol actually do more harm than the hard drugs, and I think that it is next to impossible to fight cocaine while our society pickles itself in smoke and marinates itself in alcohol. Let's do what we can to make a drug-free society, and I mean *drug*-free.

And how do we do that? I don't see any way of doing it but to change society so that the rich aren't as rich and the poor aren't as poor. Surely it can't be part of the American ideal to have a few people at one end swimming in billions, while millions at the other end have no homes and have to eat out of garbage cans.

But will we do this? Will we even try to do this?

Not as long as substantial citizens turn away from the cities and, from a safe distance, laugh and sneer at them. Not as long as we elect people who feel that helping the rich will somehow, at some future time, help the poor. Not so long as "liberal" is a dirty word, because liberals want to build a kinder, gentler society.

And what do I think the conservatives in control would say to these notions of mine? Why, I think they would "just say 'No'!" And in that case, I strongly suspect that the United States would continue to lose the fight against drugs and to lose all hope of recovery from this crisis of ours.

WARBOTS by G. Harry Stine

#5 OPERATION HIGH DRAGON   (17-159, $3.95)

Civilization is under attack! A "virus program" has been injected into America's polar-orbit military satellites by an unknown enemy. The only motive can be the preparation for attack against the free world. The source of "infection" is traced to a barren, storm-swept rock-pile in the southern Indian Ocean. Now, it is up to the forces of freedom to search out and destroy the enemy. With the aid of their robot infantry—the Warbots—the Washington Greys mount Operation High Dragon in a climactic battle for the future of the free world.

#6 THE LOST BATTALION   (17-205, $3.95)

Major Curt Carson has his orders to lead his Warbot-equipped Washington Greys in a search-and-destroy mission in the mountain jungles of Borneo. The enemy: a strongly entrenched army of Shiite Muslim guerrillas who have captured the Second Tactical Battalion, threatening them with slaughter. As allies, the Washington Greys have enlisted the Grey Lotus Battalion, a mixed-breed horde of Japanese jungle fighters. Together with their newfound allies, the small band must face swarming hordes of fanatical Shiite guerrillas in a battle that will decide the fate of Southeast Asia and the security of the free world.

#7 OPERATION IRON FIST   (17-253, $3.95)

Russia's centuries-old ambition to conquer lands along its southern border erupts in a savage show of force that pits a horde of Soviet-backed Turkish guerrillas against the freedom-loving Kurds in their homeland high in the Caucasus Mountains. At stake: the rich oil fields of the Middle East. Facing certain annihilation, the valiant Kurds turn to the robot infantry of Major Curt Carson's "Ghost Forces" for help. But the brutal Turks far outnumber Carson's desperately embattled Washington Greys, and on the blood-stained slopes of historic Mount Ararat, the high-tech warriors of tomorrow must face their most awesome challenge yet!